Programmable Logic Controllers

Programmable Logic Controllers

The industrial computer

Marco Costanzo BSc PGCE

Halesowen College, UK

A member of the Hodder Headline Group
LONDON • NEW YORK • SYDNEY • AUCKLAND

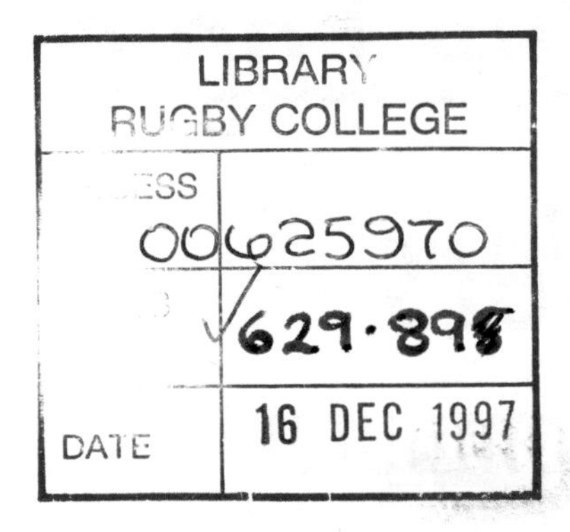

First published in Great Britain in 1997 by Arnold,
a member of the Hodder Headline Group,
338 Euston Road, London NW1 3BH
http://www.arnoldpublishers.com

British Library Cataloguing in Publication Data
A catalogue record for this book is available from the British Library

ISBN 0 340 69258 8

Publisher: Dilys Alam
Production Editor: Wendy Rooke
Production Controller: Rose James

Typeset in 11//13 pt Garamond by Saxon Graphics Ltd, Derby
Printed and bound in Great Britain by S.W. Arrowsmith Ltd, Bristol

Contents

About the author

Marco Costanzo is an incorporated electronic and systems engineer with many years experience in the design, development and application of specialized systems in the chemical, process, manufacturing and defence industries.

He has a BSc (Honours) and PGCE (FAHE) degree. Since 1992, he has been a full time lecturer at Halesowen College, West Midlands, lecturing in the Engineering Division to students on the BTEC and City and Guilds courses in electronics, microprocessesors, industrial control and engineering technology related subjects.

Preface

As the General National Vocational Qualification (GNVQ) is now becoming very popular in many subject areas of engineering, the old traditional style of the City and Guilds of London Institute (CGLI) and the Business and Technical Education Council (BTEC) courses are now being replaced by these new competence-based GNVQ criteria.

During my industrial work experience of many years in the design, development and application of systems in the chemical, process, manufacturing and defence industries, and also as a full-time lecturer, I have found that employers are constantly looking for individuals who have a multitude of skills.

These individuals will also be expected to update their skills and have the ability to contribute effectively in the workplace. The essential skills are numeracy, creativity and communications, including the ability to adapt to changing circumstances in the workplace in which sophisticated technology is playing an increasingly important role.

As technology is changing very fast, virtually everybody will have to become computer-literate and will need competencies beyond that. In other words, people will need to look at their skills in competency terms, analyse the gaps and then fill these gaps by further training.

The programmable logic controller (PLC) is part of the computer-aided engineering skills that are required in industry and the aim of this book is to provide the individual with an introduction to programmable logic controllers and their relevance to engineering.

This book offers an opportunity to identify the PLC, to derive simple programs, investigate types of communication links and identify methods of producing and storing text and documentation. The reader will also be able to investigate the methods of testing and debugging hardware and software associated with the preparation of a PLC program and then to produce and stimulate a program for a typical PLC system.

Acknowledgements

Extracts from BS 3939 are reproduced with the permission of BSI. Complete editions of the standards can be obtained by post from BSI Customer Services, 389 Chiswick High Road, London W4 4AL.

The author would like to thank:

Honeywell Control Systems Limited
Honeywell House
Arlington Business Park
Bracknell
Berkshire
RG12 1EB

for providing material on their latest technology, the 'Smart Distributed Systems' for Chapter 9.

HTE Automation Ltd
Watt House
Dudley Innovation Centre
Kingswinford
West Midlands
DY6 8XZ

for their permission to use the Data and Specifications Sheet of the GE Fanuc Series 90 Micro PLC unit.

Toshiba International (Europe) Ltd
Waterview House
1 Roundwood Avenue
Stockley Park
Uxbridge
Middlesex
UB11 1AR

for their permission to use the Data and Specifications Sheet of the Toshiba M20/40/EX100 Series PLC units.

1
Program, Plan and Production

1.1 Introduction to programmable logic controllers

There is a constant need for process control systems in the manufacturing industries to produce a better-quality product more efficiently and at a low cost. This has led to the evolution of the automated system.

Since the mid 1970s advances in electronics and especially microelectronics have made a revolution in industry from small to large systems, from sweet manufacture to a complex car plant. To ensure that the plant or machinery operates within the tolerances and at the correct speed, etc., it must be programmed and controlled.

Looking back to the early 1960s, many large industrial processes and production lines had to be changed with the development of new products coming on to the market. One of the problems associated with this was that large cabinets consisting of relays, timers and counters all had to be rewired for the new production line. This was both expensive and time consuming.

With the development of computer technology for industrial applications, the programmable logic controller was developed to replace the discrete relay, timer and counter. The programming language is based on the familiar relay symbol ladder diagram technique that electricians and technicians know and use.

The advantage of this programming technique meant that technicians were able to reprogram the production equipment in a short period of time. It was also readily accepted by many design engineers, who are now in the 1990s currently using the same basic ladder logic with advanced instructions.

When considering the installation of a process control system, the simplest arrangement should be used which is adequate for that control requirement, as well as being compatible with health and safety specifications.

The types of process control available, are as follows:

- open loop control
- closed loop control
- sequential control.

1.1.1 OPEN LOOP CONTROL

An example of open loop control is a conveyor belt driven by a motor at a constant speed so that items can be placed on to and removed from the belt by robot arms; see Fig. 1.1. If a number of items is loaded on to the conveyor belt and the motor is running at the correct speed supplied by the correct current then all is well. In practice what happens is that the load is not evenly placed on to the conveyor belt and this tends to cause the motor and the electrical supply to it to fluctuate. This occurrence cannot be predicted so the system will change its speed and is no longer under control. It is called open loop because there is no relationship between the input to the system and the output of the system, i.e. no monitoring is available.

1.1.2 CLOSED LOOP CONTROL

With this method a tachometer is placed into the system to monitor the speed of the belt. A human operator can then continually monitor the speed of the belt from the safety of a control room. If the operator observes a change in the speed of the conveyor, the electrical power to the motor is adjusted accordingly. With this system the operator closes the loop and relates input to output; see Fig. 1.2.

However, the disadvantage of this system is that no human is infallible; a loss of concentration, slow response and poor coordination could mean that the system will not operate accurately enough to maintain specific standards.

To allow the operator to perform a supervisory role instead we would incorporate an 'automatic closed loop control'; see Fig. 1.3. In this next example the tachometer is replaced by a tachogenerator which will monitor the speed of the conveyor belt and produce an electrical voltage which is proportional to the

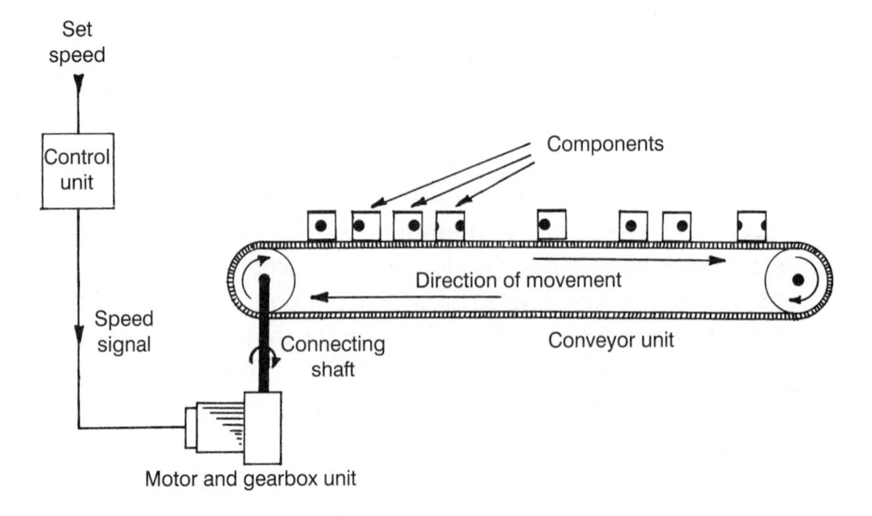

Figure 1.1 Open loop control

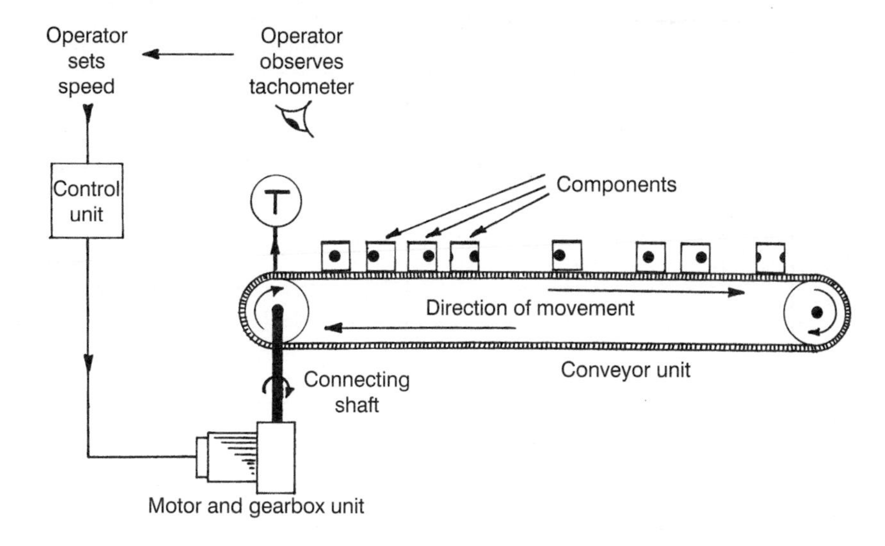

Figure 1.2 Closed loop

speed of the system. The output from the tachogenerator is the feedback signal to the control unit and this is compared with the set speed control. An error between the two voltages will change the speed of the motor.

Figure 1.4 shows the components of the control unit. It consists of an error detector (shown by the hot cross bun symbol). This error signal is then amplified and controls the speed of the motor. Looking at the diagram it can be seen why it is called a closed loop system; information regarding the speed of the motor is fed back to the amplifier which in turn controls the motor speed.

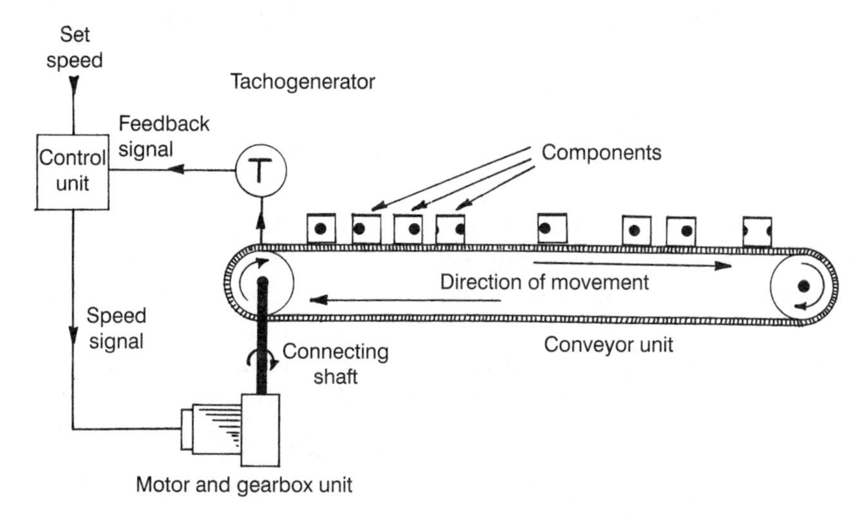

Figure 1.3 Automatic closed loop control

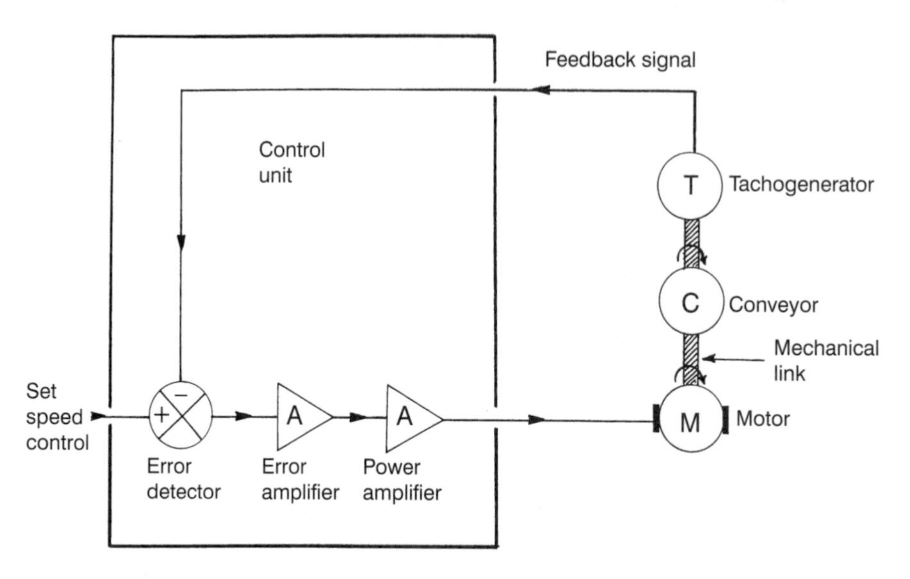

Figure 1.4 Components of the control unit

With this motor speed control system, the tachogenerator is continuously monitoring the output, and the input to the motor is being continuously updated. This continuous monitoring and updating leads us to the most important form of control in the manufacturing and process industries.

1.1.3 SEQUENTIAL CONTROL

With sequential control the aim is to ensure that the correct sequence of events takes place. The programmable logic controller (PLC) is a sequential controller; see Fig. 1.5. A sequence of instructions is programmed by the user to the PLC's memory and when this program is executed the controller will operate a system to the correct processing specifications. More complex manufacturing systems will combine both closed loop and sequential control methods to be implemented.

As an example of sequential control consider three conveyors to be turned on in the correct sequence; see Fig. 1.6. Each conveyor has a sensor which will detect whether it is moving or stationary. The conveyors must be started one at a time and the sensor from conveyor Xl will initiate the start for conveyor X2. Conveyor X3 will start some time later, initiated by the sensor from conveyor X2, and so on. The reasons for monitoring and starting the conveyors one at a time is to reduce the loading on the electrical mains supply and to make sure that they are operating at the correct speed.

The program sequence would be as follows:

Step 1 start the conveyor Xl;

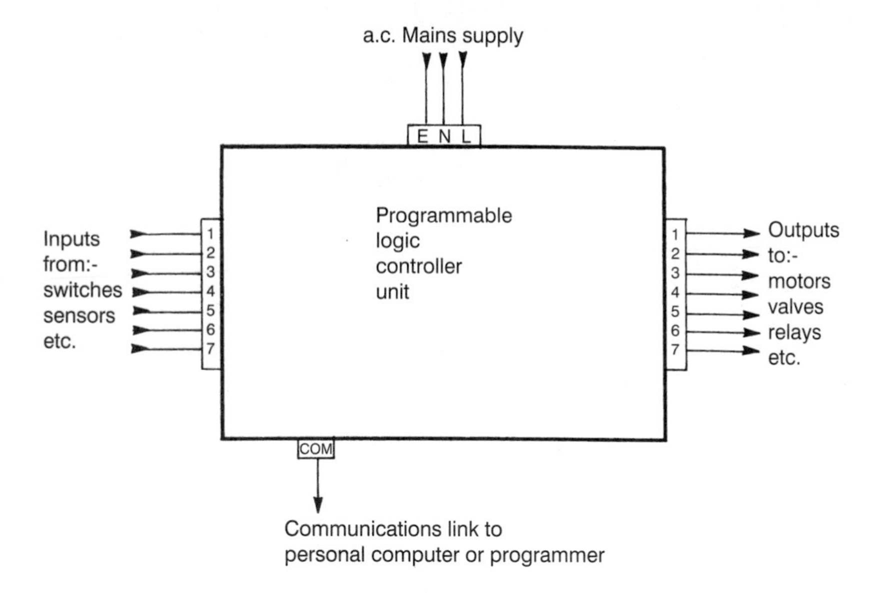

Figure 1.5 A sequential controller

Step 2 is the conveyor Xl running?
Step 3 if 'yes', start the conveyor X2 after 20 seconds;
Step 4 is the conveyor X2 running?
Step 5 if 'yes', start the conveyor X3 after 20 seconds;
Step 6 is the conveyor X3 running?
Step 7 if 'yes', continue with the program... .

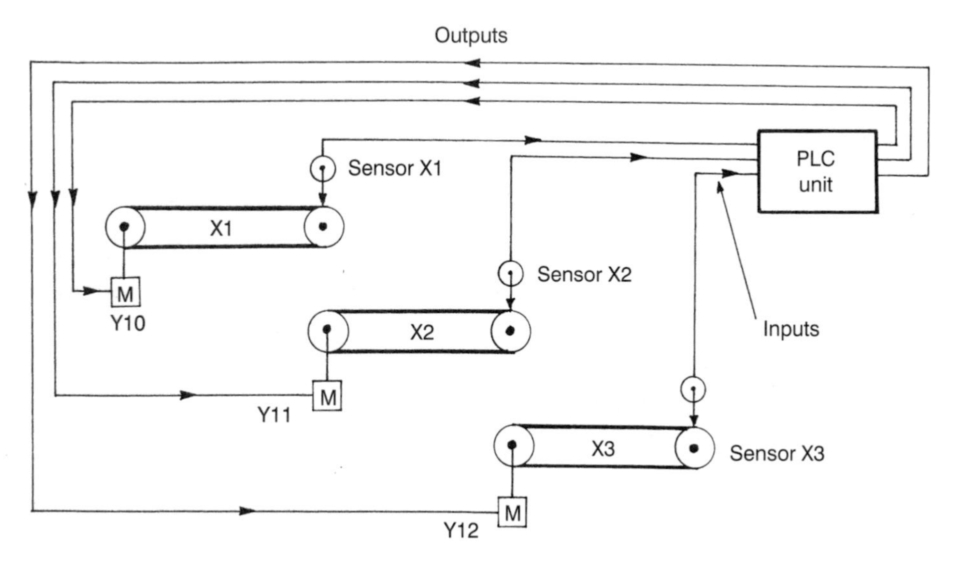

Figure 1.6 Conveyor sequencing control

In this example the conveyors will each have two sensors, one to control the motor speed and the other to detect the movement of the conveyor. Therefore, the PLC will turn all the conveyors ON in the correct sequence based on the information it receives from the sensors.

It is useful to show this sequence in a flowchart; see Fig. 1.7. We will come back to this example later to discuss the type of program required for the PLC.

1.2 The PLC unit

By now you may be wondering what is inside a PLC unit; take a look at Fig. 1.8. It can be seen that it consists of a number of modules and devices and is similar in nature to a computer.

1.2.1 POWER SUPPLY MODULE

The power supply module is used to feed all the devices of the system and the typical operating values are 24 V d.c., 110 V a.c. and 230 V a.c. It is most important that users check the power supply availability before they purchase the PLC unit.

1.2.2 CENTRAL PROCESSING UNIT

The central processing unit (CPU) is the brain of the PLC unit and consequently it is the device which controls all of the operations in the correct sequence. However, it must be understood that the CPU does not possess any intelligence as it cannot think for itself; it will only follow a set of instructions held within the memory.

1.2.3 INPUT AND OUTPUT

The inputs and the outputs (I/O) have interfacing and multiplexing circuits which are connected to the CPU via an internal bus system. PLCs will have a range of I/O; for example, 20 I/O means that there are a total of 12 inputs and a total of 8 outputs (12+8=20), but there could be a different combination for other PLC systems.

There are a number of memory devices that are used in a PLC: ROM, RAM and EEPROM. Typical memory capacities available for user program are 4K and 8K, depending on the size of the system to be controlled.

1.2.4 READ-ONLY MEMORY

Read-only memory (ROM) is programmed by the manufacturer and is a permanent non-volatile memory which stores the operating system program and data.

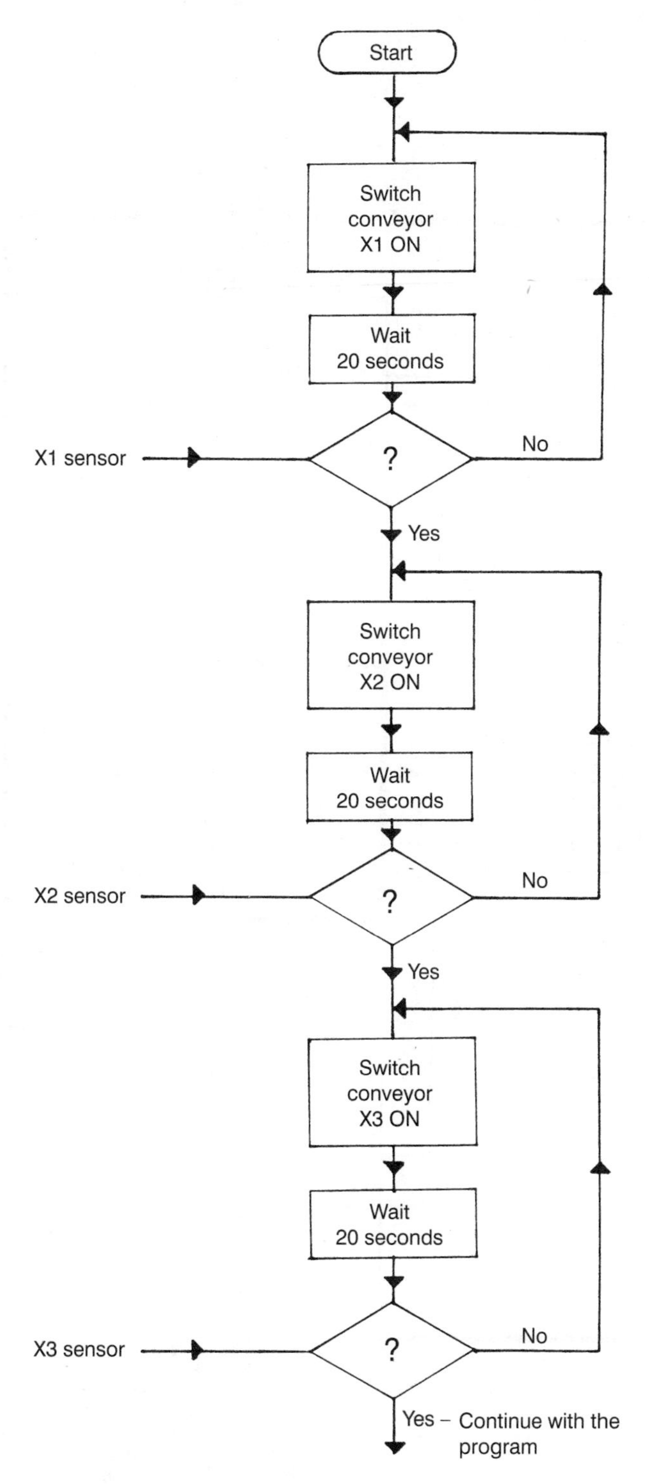

Figure 1.7 Flowchart for conveyor sequence

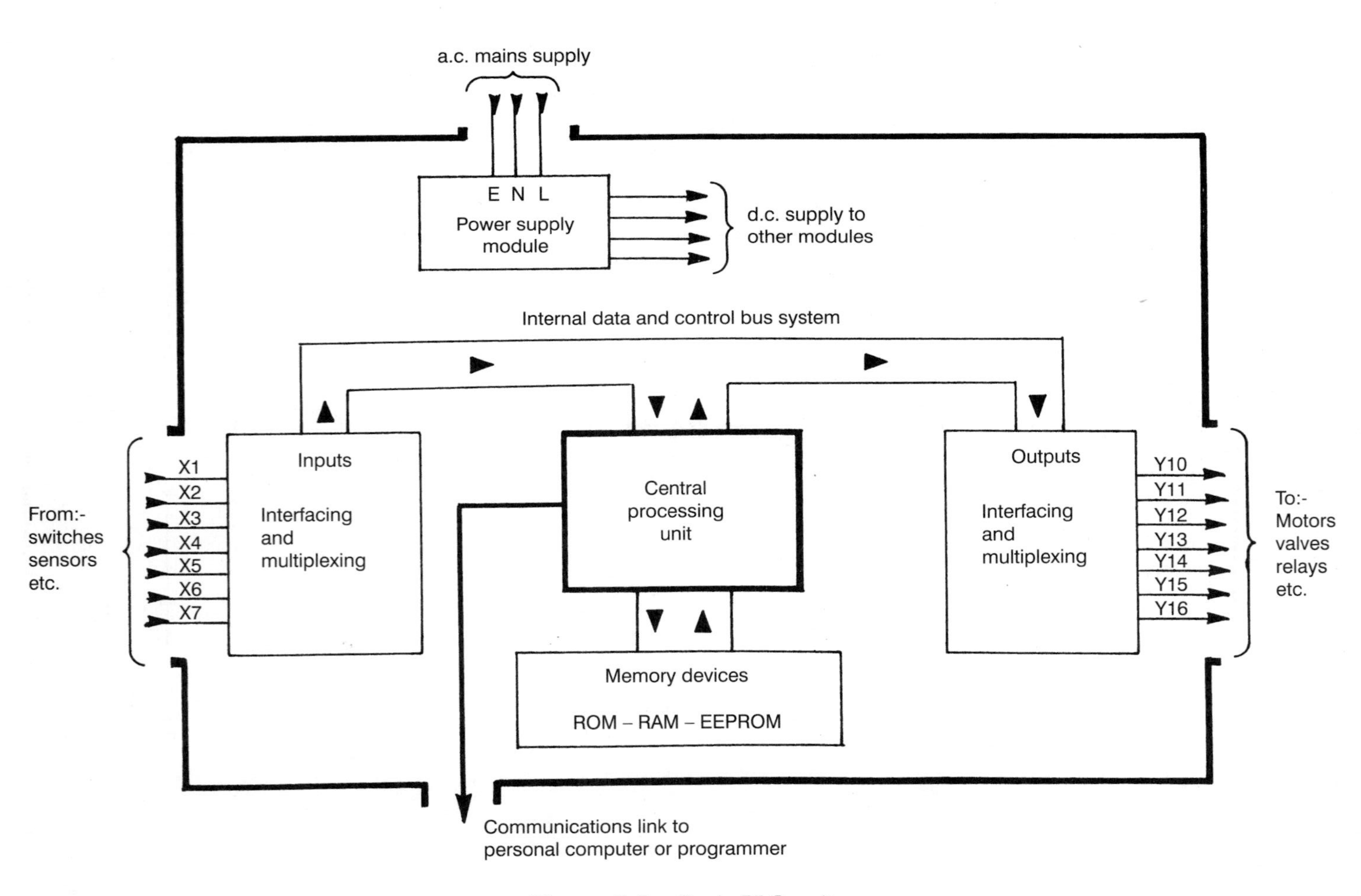

Figure 1.8 Basic PLC unit

1.2.5 RANDOM ACCESS MEMORY

Random access memory (RAM) is a read/write memory device and is volatile. The data is lost when the power is switched off. However, it is flexible as data can be modified and stored while the PLC is operating.

1.2.6 ELECTRICALLY ERASEABLE PROGRAMMABLE ROM

Electrically eraseable programmable read-only memory (EEPROM) is similar to the ROM device. The EEPROM has the advantage of being programmed by the user and in the event of a power loss will not lose its data. The EEPROM can be erased by using electrical pulses, but the writing of data into the EEPROM does take a lot longer than for example into RAM. It is normally used as a back-up device.

1.2.7 COMMUNICATION LINKS

A communications link is available to load the user program. This can be done by using either a hand-held programming keypad or a personal computer with the appropriate software package.

1.3 The PLC hardware and physical capabilities

You may now be wondering what the PLC looks like. Take a look at the specification sheets at the back of this book for typical PLCs: a unitary and a modular type. These are typical design characteristics of modern programmable logic controllers.

1.3.1 UNITARY TYPE PLC

The unitary type is self-contained, i.e. has a power supply, a CPU and a limited number of inputs and outputs. It is sometimes called a shoebox as the illustration depicts and provides an economical alternative to more expensive controllers. It is used for the control of small conveyors, palettizers, car wash, punch press, start/stop sequencing, industrial washing machines, hydraulic and pneumatic control systems and materials handling. There are many other applications too numerous to mention.

1.3.2 MODULAR TYPE PLC

The modular type, however, is a PLC which can be constructed using separate modules, i.e. power supply, CPU, inputs, outputs, timers, counters, analogue-to-digital converters, digital-to-analogue converters, expansion modules and other types of communication and network modules. This system offers greater flexibility for the user as it can be designed for a specific application such as automatic machine control and process control systems. It has a greater input/output configuration and because of its modular construction it is easily modified and can be expanded to a much larger system. The modular unit is sometimes called a rack mounted type.

1.4 The PLC operational sequence and requirements

As we shall see later the user program is executed by the PLC systems software. The speed at which the operations are carried out will depend mainly on the performance of the processor in the CPU. The usual method of operation by a PLC is to scan all the inputs, process the user program and then scan all the outputs. This is called a scan cycle, as shown in Fig. 1.9. You can see that there is a time taken for each scan. First, the input values are scanned and recorded in memory, followed by the user program and then all the outputs are updated accordingly; then the scan cycle repeats.

As mentioned earlier, the PLC is programmed by entering and storing a sequence of instructions to the PLC's memory by either a programming keypad or a personal computer. The programs can be constructed either on-line or off-line as required.

1.4.1 ON-LINE PROGRAMMING

On-line programming is implemented with direct connection to the PLC, where changes to the program can be made while the PLC is in running

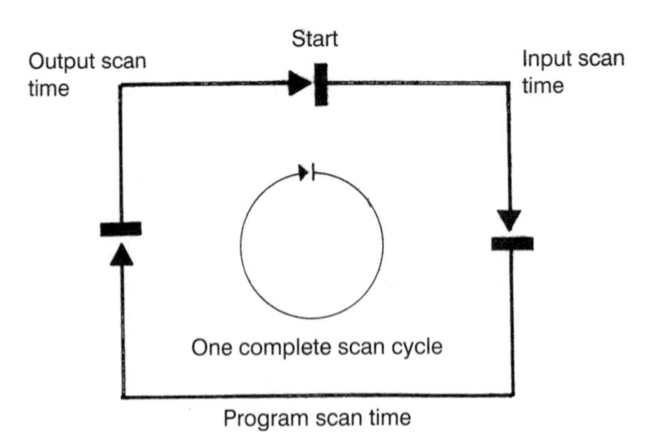

Figure 1.9 Scan cycle

mode. Timers, counters and other instruction formats can have their contents altered.

1.4.2 OFF-LINE PROGRAMMING

Off-line programming is used to construct new programs prior to running a process. This allows any programs to be checked and verified by the user and avoids the problems of any incorrect commands to the process, i.e. invalid instructions and/or input/output references. This is the preferred method of programming a PLC.

The type of programming format generally used for PLCs is ladder logic. This is a graphical representation of the hard-wired system as used by electricians and includes switches, timers, counters and many other instructions in a structured way. This leads us to an efficient way of programming and executing PLCs for process control.

1.5 The PLC program format and key codes (icons)

Ladder diagrams have been used by electricians for many years to record and document electrical circuits. When programmable logic controllers were first introduced, a simple program was required so that operators could learn and use a new program technique with minimum training. The ladder logic diagram was developed for the PLC as it was understood by all electrical and electronic engineers in the industry.

The basic ladder functions are shown in Fig. 1.10. These are the typical key codes or icons that are used and if you take a look at the specification sheets there are many other instruction codes which are used for the purpose of advanced programming techniques.

The ladder logic diagram shown in Fig. 1.11. is a typical format of many programmable logic controllers. It consists of a number of input switches and a relay coil which make up a rung of the ladder diagram. The left-hand side of the diagram is the live supply and the right-hand side is the neutral or common supply. The contacts must be active, i.e. closed, in order for the relay coil to operate. Can you see the similarity between the circuit diagram and the ladder logic diagram? Only one rung is shown, but there can be many rungs to a single program, hence the term 'ladder'. We shall see more examples of ladder logic later.

1.6 The PLC internal registers, devices and functions

There is a number of internal registers and devices used for the purpose of data transfer and arithmetic operations, as well as registers which store the on/off information of contacts and coils in ladder logic.

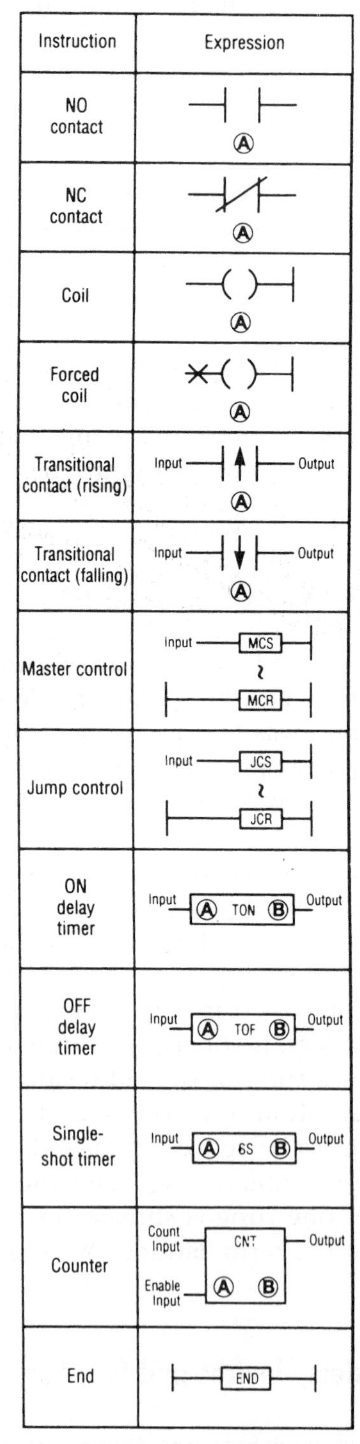

Figure 1.10 Basic ladder functions

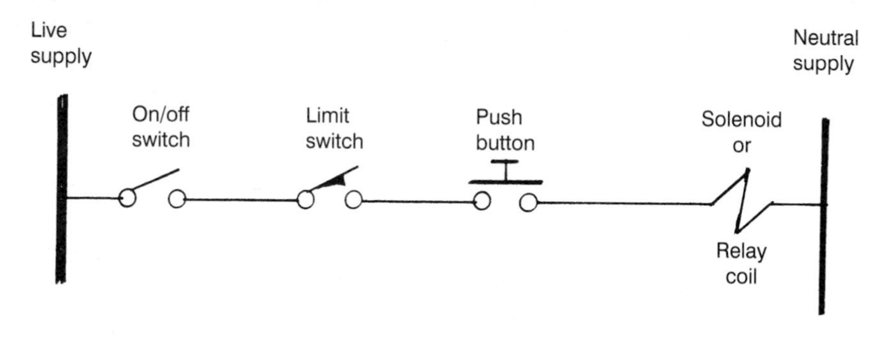

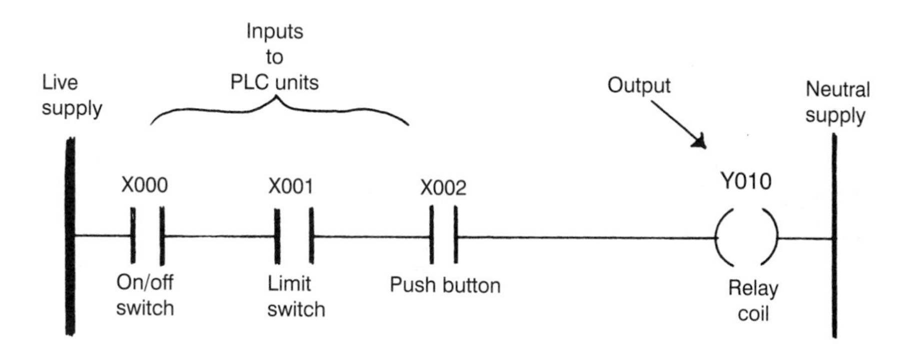

Figure 1.11 Ladder logic format

1.6.1 EXTERNAL INPUT DEVICES

The external input device (X) indicates the state of the input through the input module. These devices can be used many times in the program and the function type X is assigned to the various input modules, e.g. X000 to X009.

The coding XW is an external input register word for storing numerical information, such as the analogue inputs, i.e. address locations XW00 to XW63.

1.6.2 EXTERNAL OUTPUT DEVICES

The external output device (Y) indicates the state of the output through the output module. These devices can only be used once for the relay coils in the program; however, they can be used many times in the program as a contact. The function type Y is assigned to the various output modules, e.g. Y010 to Y019.

The coding YW is an external output register word for storing numerical information, such as the analogue outputs, i.e. address locations YW00 to YW63.

1.6.3 THE AUXILIARY RELAY DEVICE

The auxiliary relay is an internal register and is used to store intermediate results of sequences. It is usually denoted by the letter R, and would be a relay device in memory, e.g. addresses R000 to R63F. It does not exist as an output coil.

1.6.4 DATA REGISTERS

The data registers are similar to the auxiliary relay device and they cannot be used for an external device. The use of the data register is to store information and to retain data in the event of a power failure. For example, it may have addresses D0000 to D1535.

1.6.5 TIMER REGISTERS

The timer registers are used for storing the timer data and instructions, including the time remaining whilst it is being executed. For example, it may have addresses T000 to T127.

1.6.6 COUNTER REGISTERS

The counter registers are used to store the data and instructions for the counter device. For example, its addresses may be C00 to C95.

1.7 Summary and test questions

The aim of this chapter was to focus on the basic structure of the PLC and the type of control techniques that are employed. Below are some typical test questions that students must be able to answer if they are studying a PLC programming course with the CGLI or BTEC.

1. What is meant by sequential control?
2. Explain the main difference between RAM and ROM.
3. What is the advantage of using EEPROM in a PLC?
4. What is a CPU?
5. What is an auxiliary relay?
6. Explain the meaning of scan cycle.

7. Explain the meaning of I/O.
8. What is the difference between on-line and off-line programming?
9. What is the difference between a unitary type PLC and a modular type of PLC?
10. What are the main components of a typical PLC unit?

2
Ladder Logic and Truth Tables

2.1 Ladder logic functions

In the previous chapter I said that a typical ladder logic rung consisted of a number of input switches and an output relay coil. We can now look at some examples of logic circuits used with PLC programming techniques. The PLC can thus be regarded as a system with a number of inputs and an output as shown in Fig. 2.1.

Before we proceed to the theory of logic gates, it must be realized that the switches and contacts shown in Fig. 2.2 are shown in their normal condition. Hence, normally open switches or contacts have a logic '0' condition and when operated they will close to become a logic '1' condition. Similarly, normally closed switches and contacts have a logic '1' condition and when operated they will open to become a logic '0' condition.

I have also included the British Standards Institution (BSI) Symbols for each of the logic functions as an alternative system diagram.

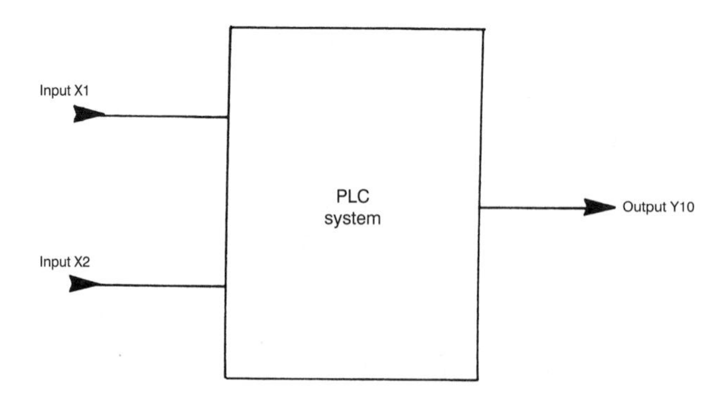

Figure 2.1 PLC system

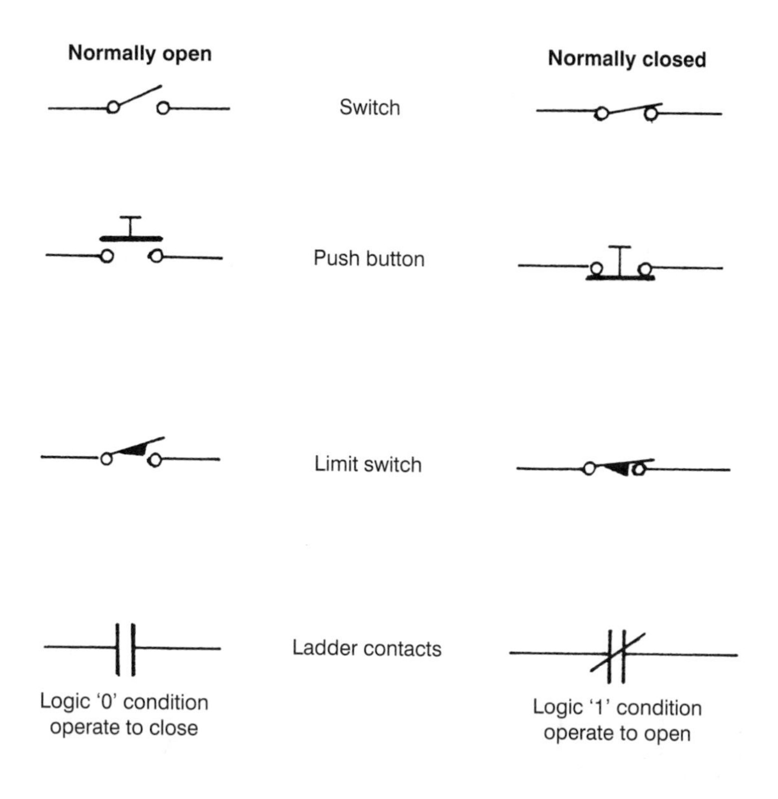

Figure 2.2 Symbols

2.1.1 THE AND GATE LOGIC

Take a look at Fig. 2.3, where the switches X1 and X2 are inputs to the system and the output is Y10. Can you see the similarity between the circuit diagram and the ladder logic program?

The circuit can best be described by the use of a **truth table**, i.e. a logic '0' represents the switch in the OFF state and a logic '1' represents the switch in the 'ON' state. The output can also be represented in the same way, i.e. a logic '0' means OFF and a logic '1' means ON; see Table 2.1.

It is clear that an output is only obtained when both the switches applied to the system are closed, i.e. they are both at logic '1'. The truth table shows all the possible combinations for the two inputs to the system.

Therefore,

X1 AND X2 = Y10.

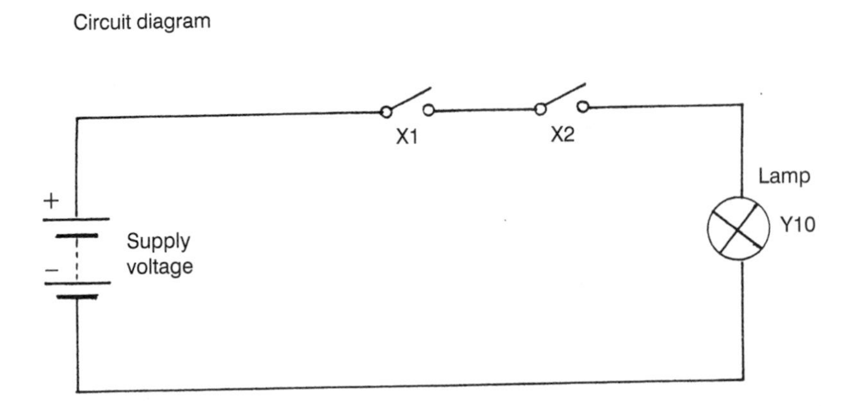

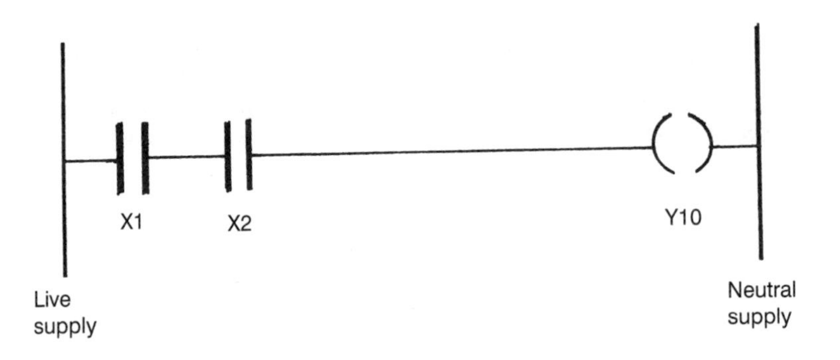

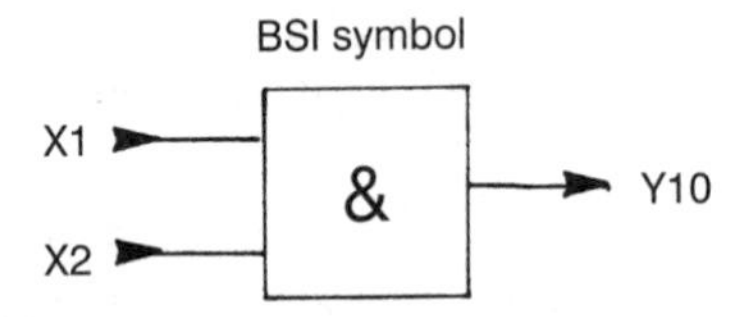

Figure 2.3 AND gate logic

2.1.2 THE OR GATE LOGIC

Figure 2.4 shows the OR gate logic function and it is clear to see that operating either switch X1 or X2 will give you an output at Y10. Again, can you see the

Table 2.1 Truth table for the AND function

Inputs		Output
X1	X2	Y10
0 (off)	0 (off)	0 (off)
0 (off)	1 (on)	0 (off)
1 (on)	0 (off)	0 (off)
1 (on)	1 (on)	1 (on)

Table 2.2 Truth table for the OR function

Inputs		Output
X1	X2	Y10
0	0	0
0	1	1
1	0	1
1	1	1*

*The condition X1 AND X2 also gives you an output and this OR gate represented by the circuit diagram and ladder logic program is called an INCLUSIVE OR.

similarity between the circuit diagram and the ladder logic program? The truth table for this function is given in Table 2.2.

In this case an output is obtained when either of the input switches are closed.

Therefore,

X1 OR X2 = Y10.

The OR gate which does not allow the X1 AND X2 condition is called an EXCLUSIVE OR gate logic.

2.1.3 THE EXCLUSIVE OR GATE LOGIC

The circuit diagram and the ladder logic program now becomes more complex in design; see Fig. 2.5. A circuit is required to break the supply to the output Y10 when the condition X1 AND X2 is operating. If you compare Fig. 2.4 and

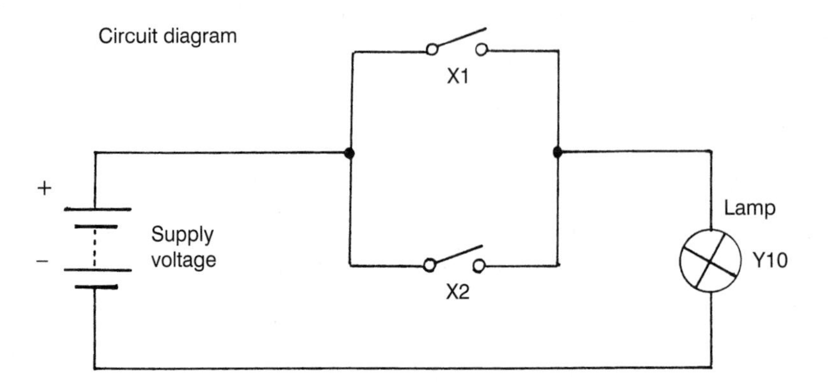

Ladder logic

X1
X2
Y10
Live supply
Neutral supply

BSI symbol

X1
X2
≥1
Y10

Figure 2.4 OR gate logic

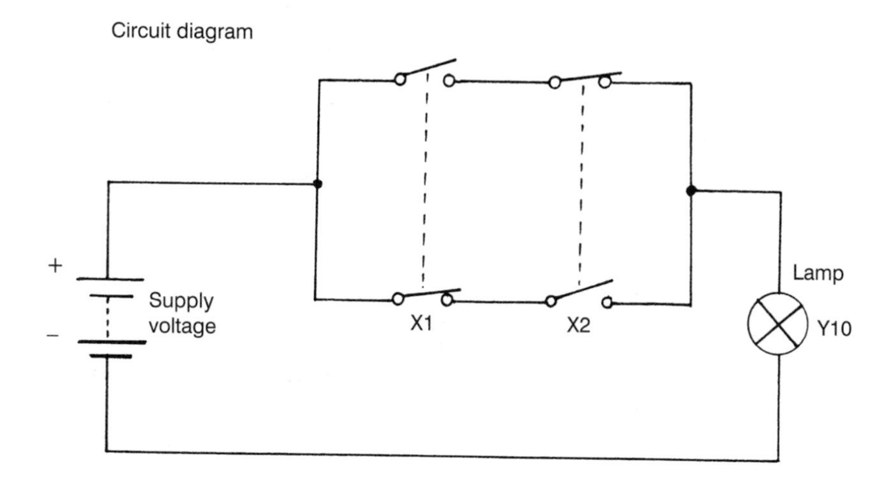

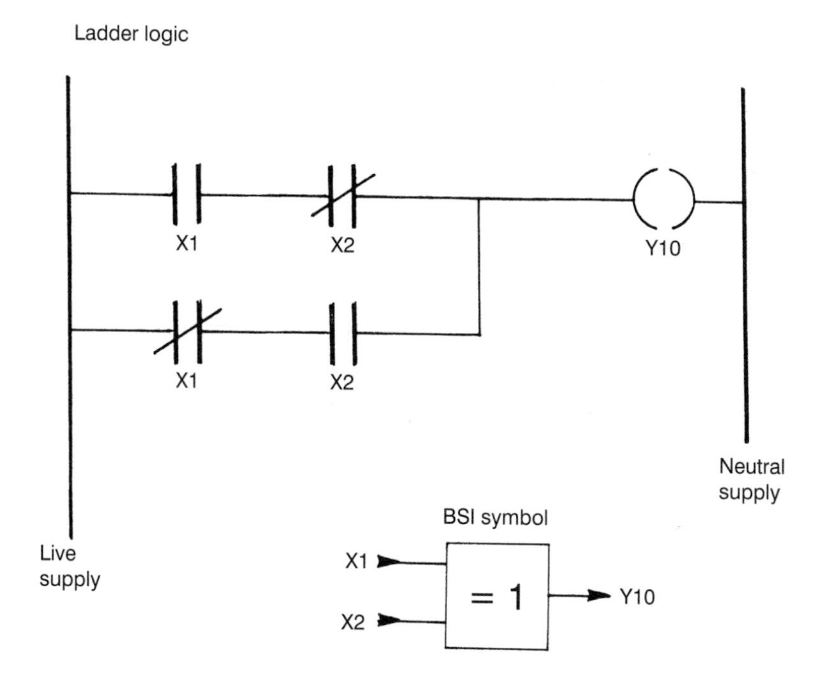

Figure 2.5 EXCLUSIVE OR gate logic

Table 2.3 Truth table for the EXCLUSIVE OR function

Inputs		Output
X1	X2	Y10
0	0	0
0	1	1
1	0	1
1	1	0

Fig. 2.5, you will see an additional circuit using normally closed switches. This is a combination of the AND and the OR gate logic functions.

Also, we are now using normally closed contacts in the ladder logic program and the inputs X1 and X2 are used a number of times. There is usually no limit as to how many times an input contact can be used in a ladder logic program or whether it is an open or a closed contact.

This circuit diagram and ladder logic program will now perform the function of the EXCLUSIVE OR, where the condition of X1 AND X2 will not allow an output (see Table 2.3).

Therfore, the EXCLUSIVE OR function can be regarded as one of non-equivalence, since the output is only obtained when the inputs are not equivalent.

2.1.4 THE NOT GATE LOGIC

The NOT gate can be represented by Fig. 2.6 using a 'press to break' switch. An output will be obtained as long as there is no input signal, i.e. the switch is not pressed; see Table 2.4.

Therefore, the NOT function can be regarded as one of inversion or negation.

Table 2.4 Truth table for the NOT function

Input	Output	
X1	Y10	
0	1	NOT X1 = Y10
1	0	X1 = NOT Y10

2.1.5 THE NAND GATE LOGIC

This circuit is an inverted AND and it can be seen from Fig. 2.7 that both of the inputs must be operated to switch OFF the output; see Table 2.5.

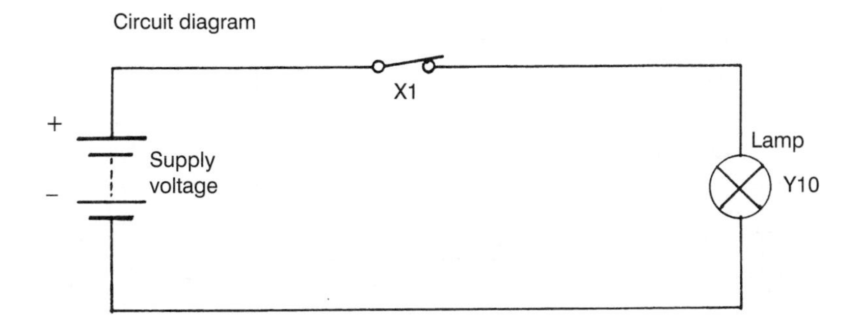

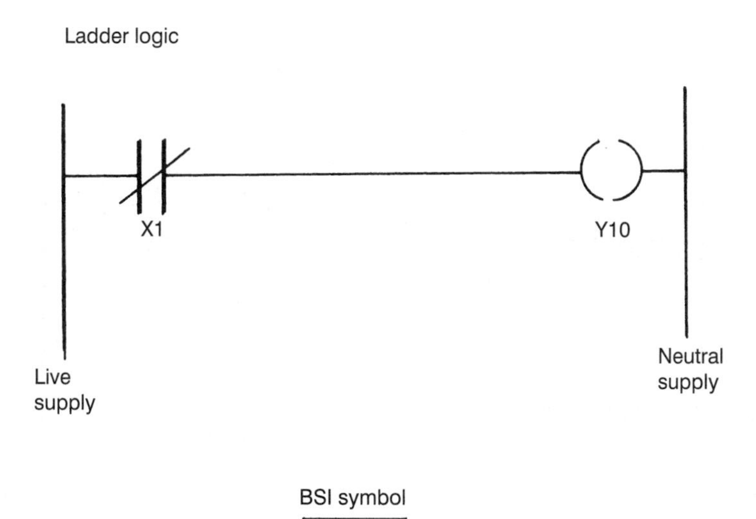

BSI symbol

X1 → 1 → Y10

Figure 2.6 NOT gate logic

2.1.6 THE NOR GATE LOGIC

This circuit is an inverted OR and it can be seen from Fig. 2.8 that if any one of the inputs are operated it will switch OFF the output; see Table 2.6.

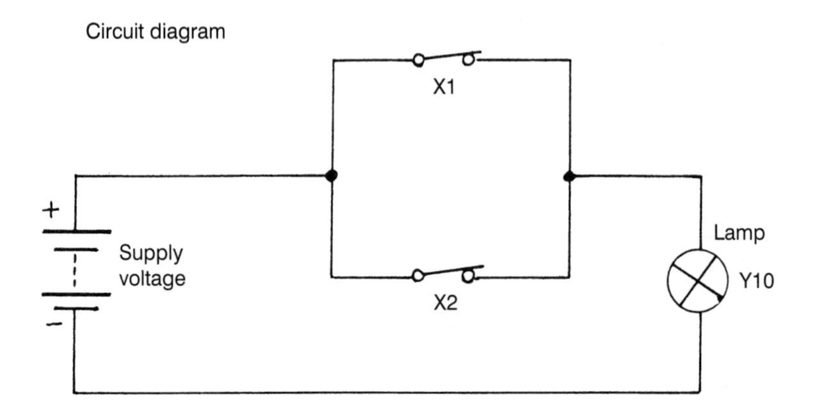

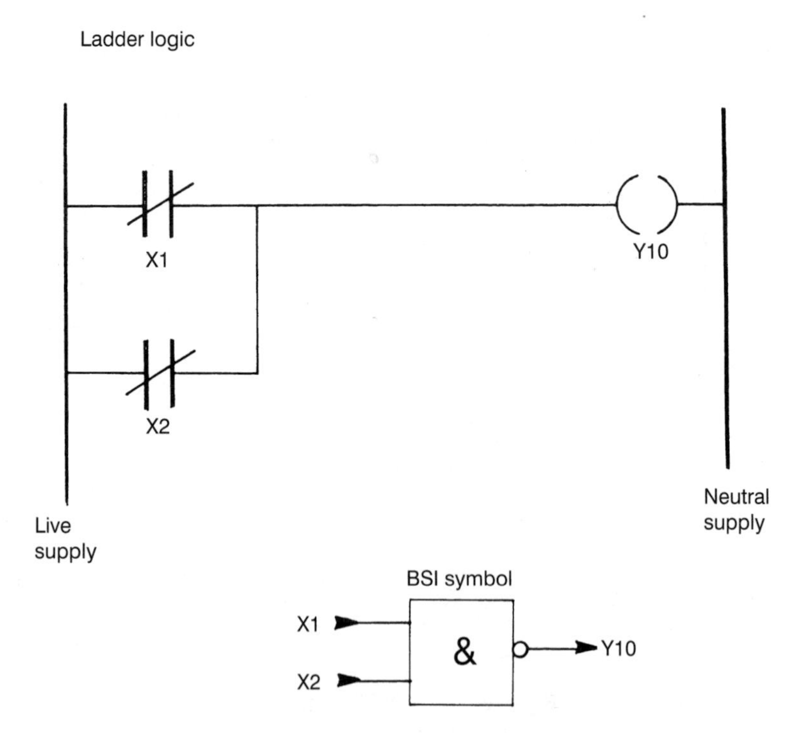

Figure 2.7 NAND gate logic

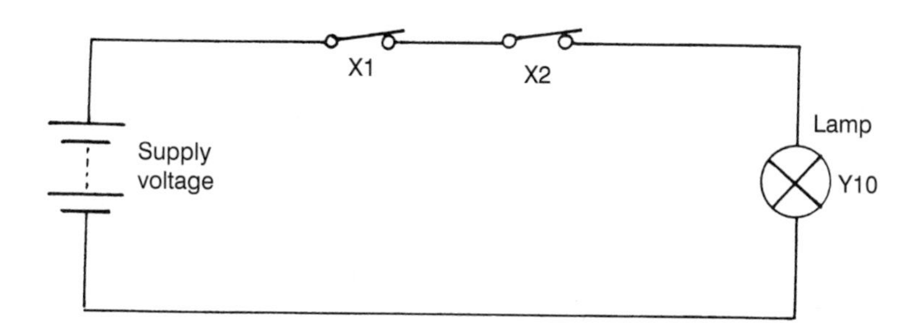

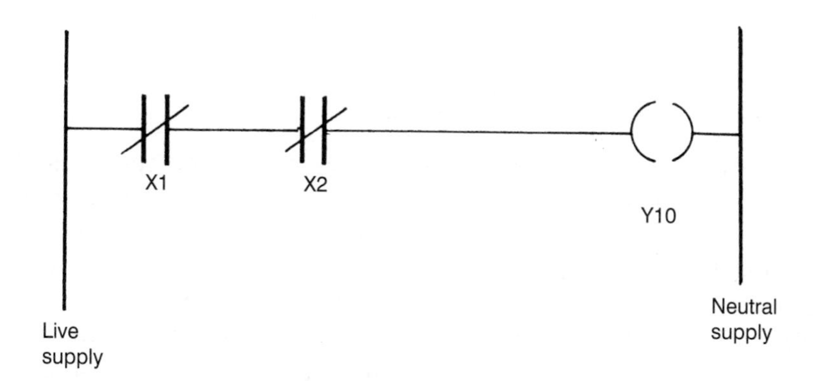

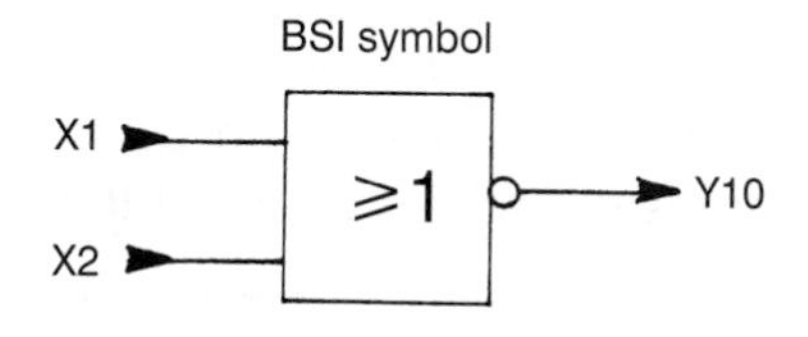

Figure 2.8 NOR gate logic

Table 2.5 Truth table for the NAND function

Inputs		Output
X1	X2	Y10
0	0	1
0	1	1
1	0	1
1	1	0

Table 2.6 Truth table for the NOR function

Inputs		Output
X1	X2	Y10
0	0	1
0	1	0
1	0	0
1	1	0

2.1.7 THE START/STOP LATCHING RELAY LOGIC

The circuit shown in Fig. 2.9 consists of a normally open start push button in series with a normally closed push button. A holding contact of the program start relay is connected in parallel across the start push button. The operation of the start push button will allow power to the relay thus closing the relay contact. When the start push button is released, the circuit will remain latched in the ON condition until the stop push button is operated to break the power to the relay.

Note that the relay R100 can have many contacts in a ladder logic program and this circuit is generally used to start and stop PLC programs.

WARNING!

This circuit must not be used as an emergency stop; see Chapter 7 for details on emergency stop circuits and applications.

2.2 Summary and test questions

The logic circuits and ladder logic program representations are the basics of PLC programming techniques and are very commonly used. Further examples will be looked at later in this book. I have only shown two inputs to the PLC system, but there can be three, four or more depending on the application. The

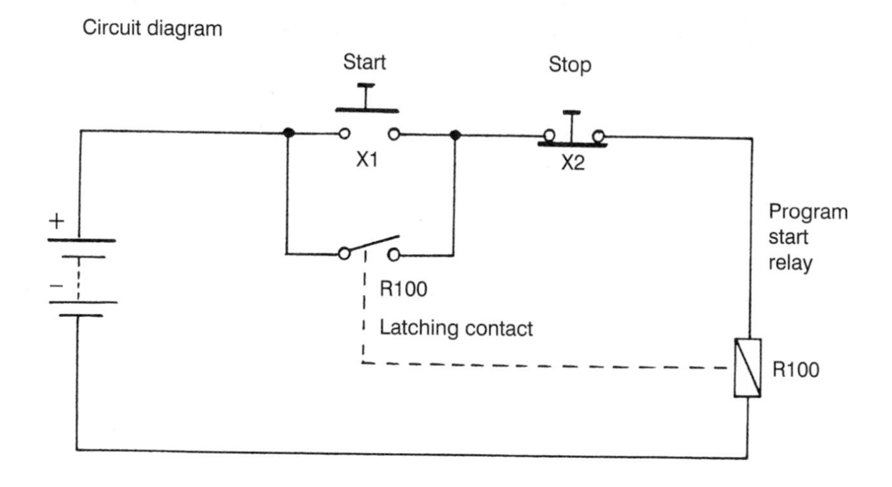

Ladder logic

X1

X2

R100

R100

Live supply

Neutral supply

Figure 2.9 START/STOP latching relay logic

limitations of the PLC system will depend on how many inputs are available for that particular PLC unit.

Test questions of what students must know:

1. Explain the difference between a circuit diagram and a ladder logic diagram.
2. Describe the term 'normally open'.
3. Describe the term 'normally closed'.

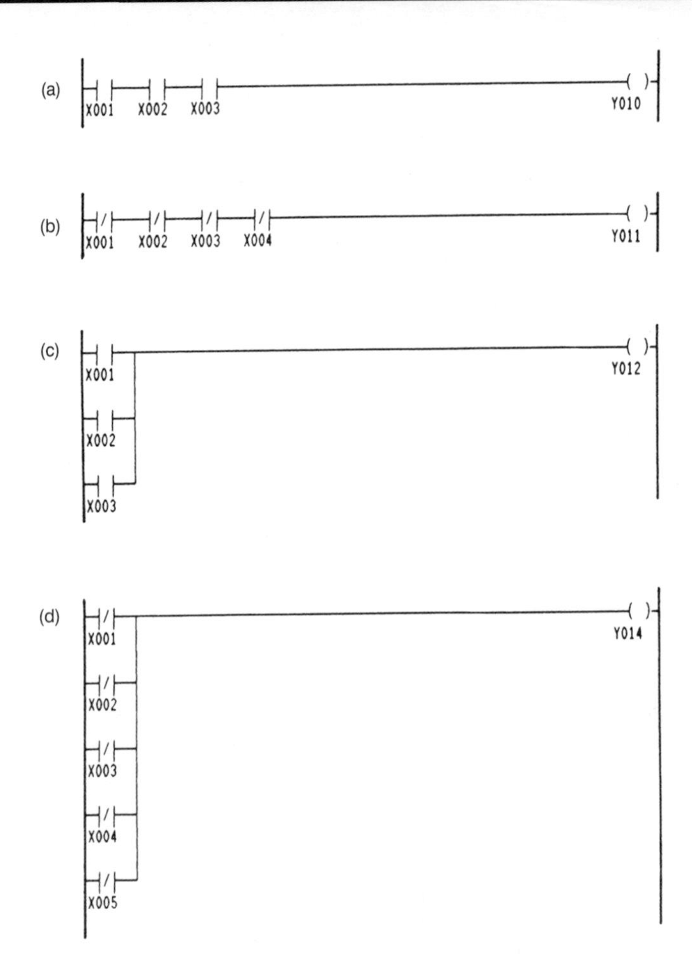

Figure 2.10 Ladder diagrams

4. Draw the AND gate ladder logic function with three inputs and one output.
5. Draw the NOR gate ladder logic function with four inputs and one output.
6. Explain the operation of an EXCLUSIVE OR gate.
7. Refer to the ladder diagrams in Fig. 2.10.
   (a) Which one is a logic OR function?
   (b) Which one is a logic AND function?
   (c) Which one is a logic NOR function?
   (d) Which one is a logic NAND function?

3
Numbering Systems

3.1 Decimal

The most commonly used system is decimal, sometimes called denary. This number system is position dependant: the further left a digit is presented the higher is its value. For example 538 can be represented as

(5 hundredths)	+	(3 tens)	+	(8 units)
(5 × 100)	+	(3 × 10)	+	(8 × 1).

3.2 Binary Logic

The operation of a digital circuit is controlled by two voltage levels. Either a low voltage, referred to as logic '0', or a high-voltage, referred to as logic '1'. High voltage is about +5 V and low voltage is less than +0.2 V. These two voltage levels are mainly used for binary logic numbering systems, which will now be discussed.

3.2.1 BINARY POSITION NUMBER

Position number	10	9	8	7	6	5	4	3	2	1	0
Denary value	1024	512	256	128	64	32	16	8	4	2	1

When we use a binary scale, we are working with a base of 2 instead of 10, so the binary number 110 becomes

Binary number	1 1 0
Position number	2 1 0
Denary value	4 + 2 + 0 = 6

We can start by looking at how we convert numbers between the two main scales of binary (more precicely, 8-4-2-1 binary) and denary.

3.2.2 BINARY-TO-DECIMAL CONVERSION

1. Write out the binary number and below each of its digits write down its proper position number, starting with position 0 at the right-hand side of the binary number and working to the left.
2. Binary is 1 0 1 1 1 0 1 0
3. Position is 7 6 5 4 3 2 1 0
4. Then consult Table 3.1, which allocates each position number to its own denary value equivalent.

Mathematically, each position number is doubled, as shown in the denary column and the table is endless.

All that is required now is to look for all the 1's (ignoring all the 0's). Table 3.2 shows the position number of all these 1's and in a separate column, the denary values. Add up the latter column and you have the denary equivalent of your binary number.

Table 3.1

Position number	Denary value
0	1
1	2
2	4
3	8
4	16
5	32
6	64
7	128
8	256
9	512
10	1024
11	2048
	etc.

Table 3.2

Position number	Denary value
1	2
3	8
4	16
5	32
7	128
	Total = 186

Table 3.3

437/2 =	218	$r = 1$
218/2 =	109	$r = 0$
109/2 =	54	$r = 1$
54/2 =	27	$r = 0$
27/2 =	13	$r = 1$
13/2 =	6	$r = 1$
6/2 =	3	$r = 0$
3/2 =	1	$r = 1$
1/2 =	0	$r = 1$

3.2.3 DECIMAL-TO-BINARY CONVERSION

1. Write down the number you wish to convert. Say it is 875.
2. Divide it by 2; write the result below and the remainder (0 or 1) at the side, i.e.

$$875 / 2 = 437 \quad r = 1$$

3. Now divide the result by 2 again, once more placing the result below and the remainder at the side. Go on doing this until the last possible figure has been divided, leaving zero and a remainder of 1; see Table 3.3.
4. Now read off the remainders in order from the foot of the column upwards and you get the binary number.
   In the example given above, the binary equivalent of 875 is

1 1 0 1 1 0 1 0 1 1
MSB LSB

where MSB is called the most significant bit and LSB is called the least significant bit.

3.2.4 BINARY CODED DECIMAL

When numbers need to be continually displayed in their decimal form, whilst being processed in binary, the commonly used code is known as the binary coded decimal (BCD) system. Each denary digit of a number is represented by a 4-bit binary equivalent. The BCD-to-decimal equivalents are listed in Table 3.4. Thus the decimal number 875 would be represented by

1 0 0 0	0 1 1 1	0 1 0 1
8	7	5

In practice, the interchange between the two forms could be performed by encoder and decoder circuits which will be discussed later in this book.

Table 3.4

Decimal	Binary coded decimal
0	0000
1	0001
2	0010
3	0011
4	0100
5	0101
6	0110
7	0111
8	1000
9	1001

3.2.5 HEXADECIMAL CODING

Most modern computers, as well as PLCs, use an 8-bit (1-byte) word and will divide exactly into 4 bits (a nibble), which can be coded into a suitable shorthand called hexadecimal. A standard has been agreed that uses the first capital letters of the alphabet as shown in Table 3.5.

Table 3.5

Decimal	Binary	Hexadecimal digit	Exponent value	
			1	0
0	0000	0	0	0
1	0001	1	16	1
2	0010	2	32	2
3	0011	3	48	3
4	0100	4	64	4
5	0101	5	80	5
6	0110	6	96	6
7	0111	7	112	7
8	1000	8	128	8
9	1001	9	144	9
10	1010	A	160	10
11	1011	B	176	11
12	1100	C	192	12
13	1101	D	208	13
14	1110	E	224	14
15	1111	F	240	15

To convert from binary into hexadecimal coding, each byte is divided into two groups of four bits and each group is then replaced by the character shown in the table. Thus

1110, 0110 becomes E, 6.

The weighting of each character and its hexadecimal equivalent are shown in Table 3.5 for up to two character hexadecimal numbers, i.e. the exponent value. Thus

$$\begin{aligned} 6F &= 96 + 15 \\ &= 111 \end{aligned}$$

and

$$\begin{aligned} B5 &= 176 + 5 \\ &= 181. \end{aligned}$$

3.2.6 OCTAL CODING

If a binary word is divided into groups of three, each group will represent a number and these numbers are shown in Table 3.6. It is similar to the BCD table, but note the change after seven counts.

Table 3.6

Binary	Octal
000	0
001	1
010	2
011	3
100	4
101	5
110	6
111	7
001 000	10
001 001	11
001 010	12
001 011	13
001 100	14
001 101	15
001 110	16
001 111	17
010 000	20
011 000	30
100 000	40
001 000 000	100
010 000 000	200
	etc.

From this table, the binary number 110, 011, 001 can be represented as 631.

To distinguish such an octal number from binary, decimal or hexadecimal, it is common to add a suffix after the number. Thus

631_8 is octal
110_2 is binary
120_{16} is hexadecimal
101_{10} is denary.

With octal coding, the number of bits for data entry is reduced and the PLC can easily be programmed to convert the data into binary.

3.3 Summary and test questions

A knowledge of numbering systems is essential for the understanding of PLCs as well as computers. Manufacturers use coding systems extensively to specify inputs, outputs, timers, counters, etc., hexadecimal being very popular. Also, the reader will need to know the basics of numbering systems to be able to program using advanced ladder logic function instructions.

Finally, I must emphasize that PLCs can only handle binary coding. Therefore, if you use BCD, hexadecimal or octal coding for the input word, you must first convert it to binary, using the appropriate advanced instruction set, so that the internal processing capabilities of the PLC can take place. It is also possible to convert binary coding to another code for the output word.

The advanced instructions will be discussed in the next chapter.

Test yourself with the following questions.

1. Convert the binary number 11001010 to decimal.
2. Convert the decimal number 469 to binary.
3. Evaluate the hexadecimal number C6 in denary and binary.
4. What is the denary value of 1245 in BCD code?
5. Convert the octal code of 101, 101 to denary.
6. The binary output of an ADC card of a PLC is 01111111, which is equivalent to 5 V d.c. What is the value in denary:
   (a) 100
   (b) 227
   (c) 157
   (d) 127.
7. Convert the BCD code 1000, 0111, 0101 to denary. Which one is correct:
   (a) 775
   (b) 675
   (c) 875
   (d) 587.

8. What is the value of the hexadecimal number A5 in denary? Choose one of the values below:
   (a) 265
   (b) 651
   (c) 516
   (d) 165.

4
How to Write and Prove a Ladder Program

4.1 Software functions

A PLC ladder program is usually produced by using a computer with a menu-driven software program. This means that you pick the icons from the screen menu to indicate the task you want. Each selection is done by pressing a function key on the keyboard, i.e. F1–F12.

There are also the alphanumeric keys labelled with letters, numbers and other symbols, which are used for data entry. The cursor keys will move the symbols around the screen and locate data.

The cursor is the main indicator and this will give you the current location for data entry from the keyboard to the screen. The cursor is usually a flashing block and can be controlled by the [PAGE UP], [PAGE DOWN], [HOME] and [END] keys. Most PCs will combine the cursor keys with a numeric keypad, but you must first turn off [NUM LOCK] in order to use these cursor keys.

To use the function keys properly, you first need to know the label of each one; usually a menu on the screen will be displayed along with the function key number. Some PLC manufacturers may provide their software in a more elaborate form, e.g. by giving a more complete description of each function.

4.1.1 ON-LINE AND OFF-LINE OPERATIONS

It is important to understand the difference between 'On-line' and 'Off-line' programming mode. It is the difference between saving programs to the PLC memory or to the disk file.

However, you can transfer from one medium to another, e.g. save from the PLC memory to a disk file. As long as you know which mode you are working in, you can add, modify or delete to the ladder program at any stage of the development. As a guide try and remember the following:

- On-line: PLC program is stored in the memory of the PLC unit;
- Off-line: PLC program is stored on either a floppy disk or hard disk file.

4.1.2 HELP FACILITIES

By pressing a certain function key, you can call up a Help menu on to the screen to give you more information regarding the ladder logic operations. When you have finished with the help menu you can then return to the main program by pressing another function key. Note that the operation of the function keys will be different for each manufacturer's PLC software program.

4.1.3 DEVELOPMENT OF PROGRAMS

There is a number of tasks that you must do in the development of a ladder logic program. They are:

- Analyse the control requirements of the system and list all of the inputs and outputs that will be used.
- Give each input and output a device address, for example, 'START SWITCH = X000'.
- Sketch a ladder diagram, on paper, of your intended design and ideas, as this will help you write the program using the keyboard.
- When all of the symbols have been entered and you have checked the program for errors, edit the program making any necessary adjustments.
- Run the program and monitor the outcome by either simulation or a test run on the real machine if it is possible. You can then make further modifications and adjustments to the program to make sure that everything is working smoothly.
- Save the ladder program to a disk file and to the EEPROM as a back-up for the PLC.
- Prepare all necessary documentation by printing out a hard copy for reference and maintenance.

The tasks outlined above may take a few hours, a few days or even a few weeks, depending on the size and complexity of the ladder program.

4.1.4 WRITING PROGRAMS

When you start to write your ladder logic program, you must give each device an address code and in the case of a timer and a counter, etc., you must also include their preset values.

Also, when the device has been entered the cursor will automatically move to the next device location. If you make a mistake, move the cursor to the device location and either delete or insert a new symbol.

In Chapter 1 I briefly mentioned some ladder logic programming rules. It is important to remember them as they must be adhered to while programming.

- The power will flow from the left-hand side bus rail to the right-hand side bus rail.
- Each rung must start with a contact from the left and end with a relay or output coil on the right.
- Coils must not be connected to the left-hand bus rail.
- Contacts cannot be connected to the right-hand bus rail.
- Only one relay or coil may be placed on a single rung.
- Each relay or coil can only be used once in a program. If two or more have the same assignment number then an error will occur.
- Contacts can be used with the same coding many times. There is no limit other than memory space.
- Invalid entries will not be accepted. If you make a mistake the device must be entered again.
- The HALT/RUN switch on the PLC must be in the HALT mode to enter new programs or edit existing programs.
- You can use as many pages as required when writing your program as long as you stay within the memory requirements of the PLC.
- To conserve memory space it is important to place the devices from the left-hand side; see Fig. 4.1. Also, make sure that your program starts at the top of each screen, as a single rung at the bottom of a screen will waste a lot of valuable memory space.

4.1.5 FLOWCHARTS

In chapter 1, I showed you a flowchart for the operation of a number of conveyors starting up in the correct sequence. It is very common to use flowcharts

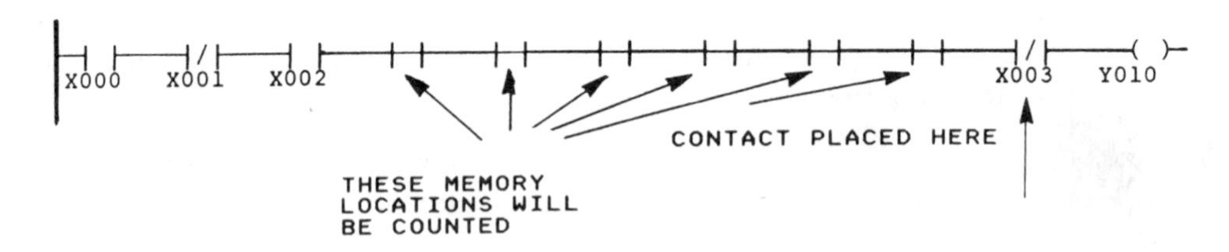

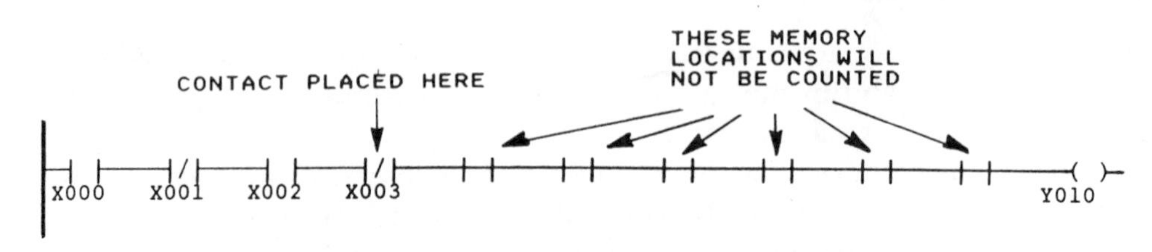

Figure 4.1 Ladder logic memory allocation

prior to programming the PLC as it will give you a visual description of the control sequence.

Some of the symbols that are used for the construction of a flowchart are shown in Fig. 4.2. They can also be used to show the branching to other subroutines of the process to be programmed. For example, Fig. 4.3 is a continuation from Fig. 1.7 in Chapter 1 and it shows a number of subroutines that could be included.

A PLC ladder program can be seen in Chapter 8 for the basic three-stage conveyor system, with these subroutines included.

4.1.6 I/O ASSIGNMENT LIST

It is useful to list all of the inputs and outputs in a table, with a description of their function. For example, the three-stage conveyor system I discussed in Chapter 1 would look like Table 4.1.

The list is not complete as you will see later. However, it is standard practice to assign inputs and outputs in this way, as it will make the programming tasks ahead easier. It also provides you with a format for the system documentation.

4.1.7 I/O CROSS-REFERENCE LIST

This is another useful map option for your ladder program documentation. The format may take many forms, but a general format may look like Table 4.2.

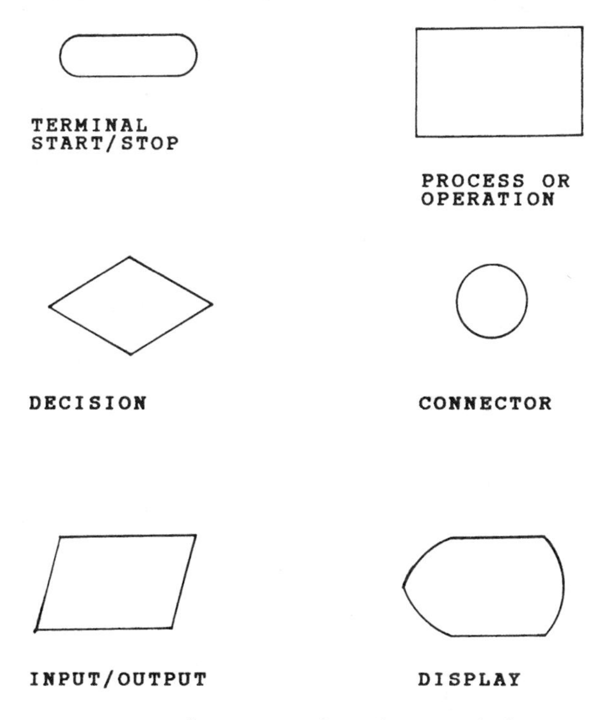

Figure 4.2 Basic flowchart symbols

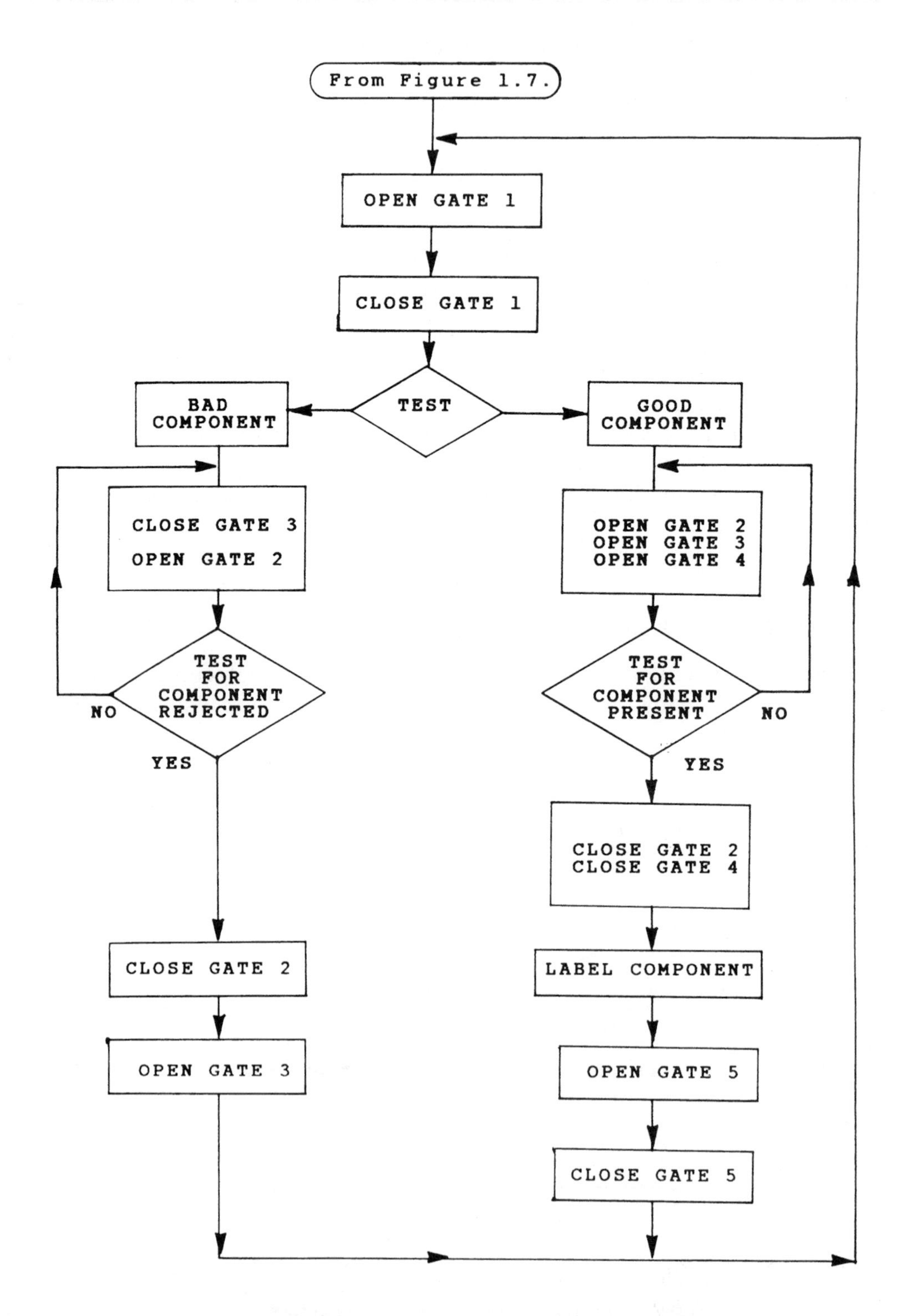

Figure 4.3 Flowchart showing subroutines

Table 4.1

PLC inputs	Sensors
X001	Conveyor No.1 movement sensor
X002	Conveyor No.2 movement sensor
X003	Conveyor No.3 movement sensor
PLC outputs	**Devices**
Y010	Conveyor motor No.1
Y011	Conveyor motor No.2
Y012	Conveyor motor No.3

Table 4.2

Inputs	Cross-reference list	
	P L T	P L T
X001	001–03–NO	002–10–NC
X002	001–04–NO	002–11–NC
X003	001–05–NO	002–12–NC

where P = Page; L = Line; T = Type

From the example in Table 4.2, X001 is shown as being used as a 'normally open' contact on page 1, line 3, and as a 'normally closed' contact on page 2, line 10. This list is generated by the PLC software on demand and again the cross-reference format may vary with different PLC manufacturers.

4.2 Timers

Timers are used extensively in PLC programming, because mechanical devices are inherently slower in their response as compared to software and electronic devices. Therefore, time delays are required within the program to synchronize the software with the mechanical devices.

Timers are also useful to delay the actions of certain devices as previously mentioned, e.g. the three-stage conveyor system, in order to reduce the surge on the electrical mains supply.

4.2.1 TIMER ON-DELAY

Refer to Fig. 4.4. A timer on-delay (TON T001) turns ON the output (Y010) when the preset time has elapsed after the input signal (X001 normally open

```
X001  {00010  TON  T001 }  ( )- Y010
PRE-SET TIME      TIMER IDENTIFICATION
```

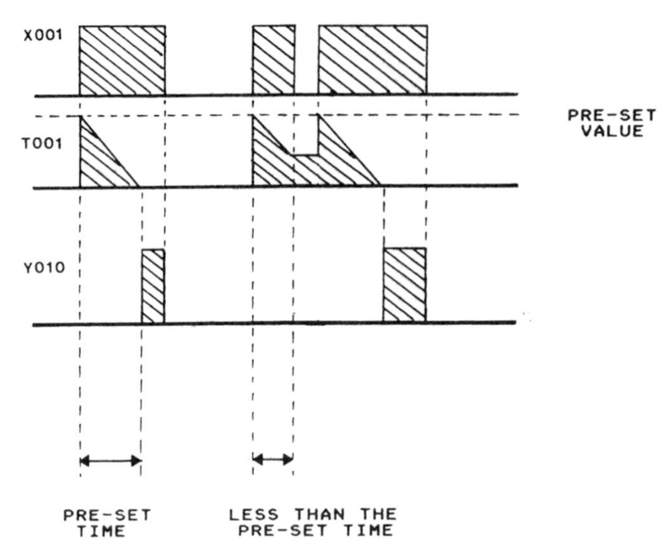

Figure 4.4 On-delay timer function

contact) comes ON. If the input signal goes OFF during the preset time, the timer will stop and the next time the input signal goes ON the timer will start to count the preset time again. As long as the input remains ON the timer will turn ON the output after the preset time.

4.2.2 TIMER OFF-DELAY

Refer to Fig. 4.5. A timer off-delay (TOF T002) will turn OFF the output (Y012) when the preset time has elapsed after the input signal (X002 normally open contact) goes OFF. If the input signal goes ON during the preset time the timer will stop and the output will remain ON. As soon as the input goes OFF again the timer will turn the output OFF after the preset time.

For example, the sequence of three conveyors must start up with a 10 s delay between each start-up, when a start button is pressed. The ladder logic program will take the form as shown in Fig. 4.6.

Task 1. Write a program to switch off the conveyors in the reverse order with a 10 s delay between each shut-down, when the stop button has been pressed.

```
X002 —{00010  TOF  T002}— ( ) Y012
```

PRE-SET TIME

TIMER IDENTIFICATION

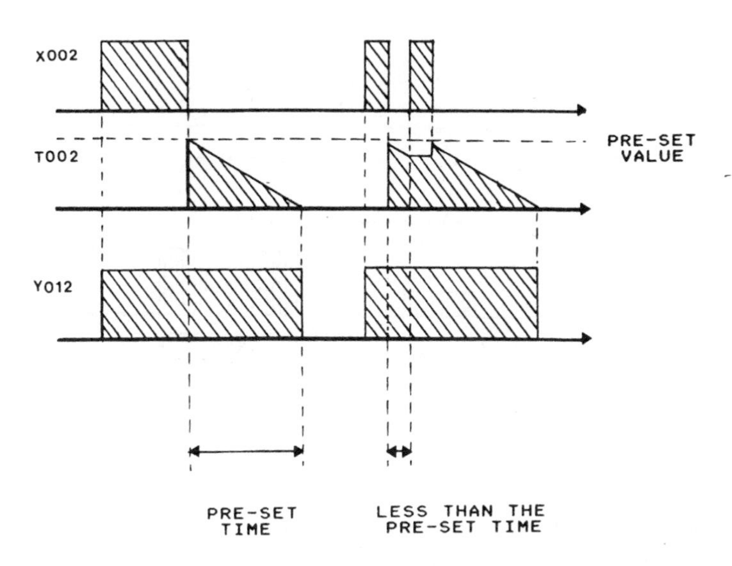

Figure 4.5 Off-delay timer function

You will need to modify the program in Fig. 4.6. The solution is at the back of the book, but have a go before you look.

4.2.3. COUNTERS

Counters are just as important as timers in the ladder logic program. They have the ability to count how many times a process has operated and they can also

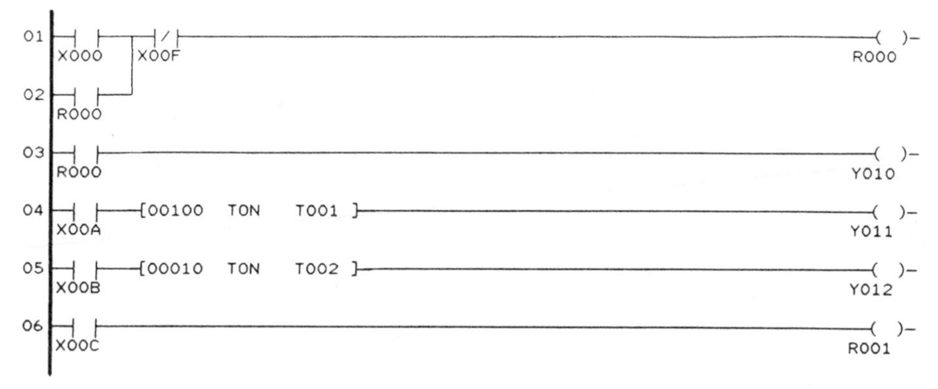

Figure 4.6 Three-stage conveyor control

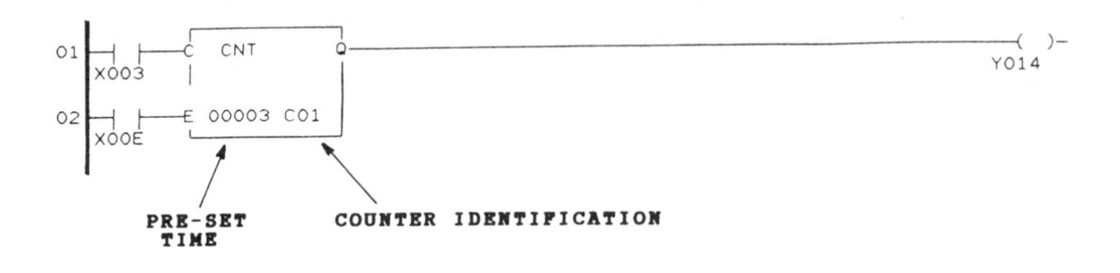

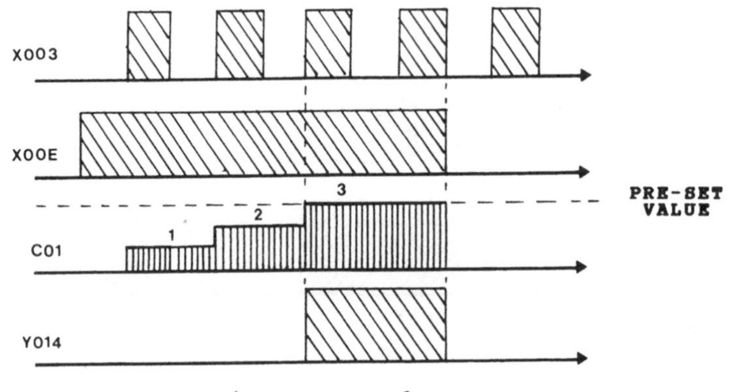

Figure 4.7 Counter function

turn outputs both ON and OFF. Figure 4.7 shows a sample program of a counter function.

The counter has two inputs, one for enable (X00E) and another for counting the number of times the input (X003 normally open contact) changes state. The preset count is stored in the counter register (C01) and when the input count reaches the preset value of 3, the counter output (Q) is turned ON, thus switching ON the output coil (Y014). When the enable input is turned OFF, the counter will reset to zero and the output is turned OFF.

Task 2. Write a program to open and close a door 10 times with an interval of 5 s for each operation. At the end of the cycle the program should stop and wait for another cycle to start again. The solution is at the back of the book.

4.2.4 REGISTERS

In Chapter 1 I said that the purpose of registers was to store and transfer word data for arithmetic and special instructions and operations. These registers can be classified as follows:

- XW: external input registers
- YW: external output registers
- RW: auxiliary relay registers
- ZW: link registers
- D: data registers

T: timer registers
C: counter registers
S: shift registers.

As an example, take a look at Fig. 4.8 which shows a ladder logic program to calculate the temperature from degrees Centigrade to degrees Fahrenheit. The mathematical formula is

$$\text{Deg. F} = \text{Deg. C} \times 9 / 5 + 32.$$

1. Line 1 converts the BCD1 input XW00 to binary and stores the value in the register of D0100.
2. Line 2 multiplies the contents of the register D0100 by 9 and stores the resultant value in the register D0101.
3. Line 3 divides the contents of the register D0101 by 5 and stores the resultant value in the register D0103.
4. Line 4 adds the contents of the register D0103 with 32 and stores the resultant value in the register D0105.
5. Line 5 converts the binary contents of the register D0105 to BCD1 and stores the value in the register D0106.
6. Line 6 inverts the contents of the register D0106 to a negative logic (NOT) to drive the output of the YW01 display.

Task 3. Write a program to convert from degrees Fahrenheit to degrees Centigrade. You will need to transpose the formula or simply reverse the actions of the program. A solution will be found at the back of the book.

4.2.5 SHIFT REGISTERS

The shift register function can be used to control the operations of a number of working stations along a production line. To amplify this statement take a

```
01├[XW00     BIN  D0100]

02├[D0100   * .   00009    > >    D0101]

03├[D0101   / .   00005     >     D0103]

04├[D0103   + .   00032     >     D0105]

05├[D0105  BCD1   D0106]

06├[D0106   NOT   YW01 ]

07├┤ END ├
```

Figure 4.8 Centigrade to Fahrenheit conversion

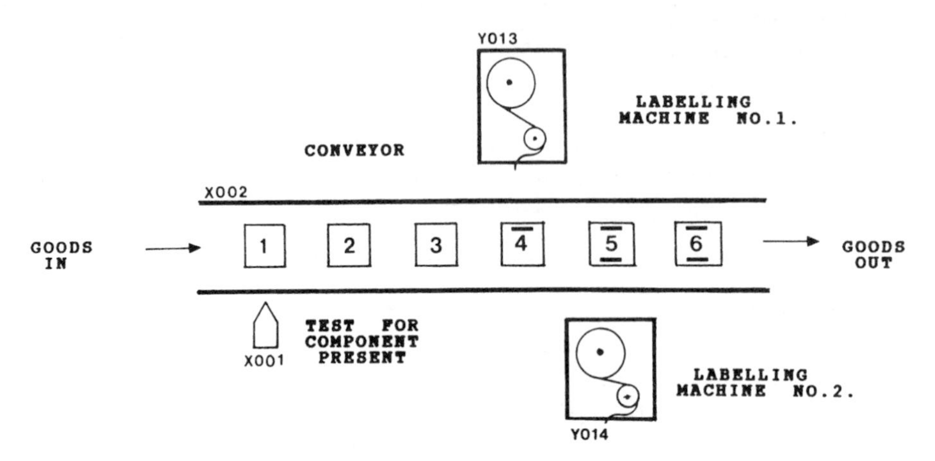

Figure 4.9 Labelling operations

look at Fig. 4.9. The goods are to be labelled in a number of ways at the end of a production run. The shift register can control the operations of the two labelling machines as the production line is synchronized with the shift register, indicated by the numerical location.

The test position enters the data into the shift register as a '1' or a '0'. A logic '1' means that there is a component present.

The action of the conveyor moving one step provides the shift register with a clock signal which increments the shift register to shift all of the data from left to right, forming a queue which provides first-in-first-out (FIFO) operation; see Fig. 4.10.

As the conveyor moves, the sensor tests for a component and inputs this information into the shift register. The contents of the shift register will now reflect the components on the conveyor and when the component arrives at the labelling station an output from the shift register will allow that machine to operate.

A logic 'O' being shifted through the register means that there are no components present, therefore the labelling machines will not operate.

Figure 4.11 shows a ladder logic program for the labelling stations. Note that the number shown in brackets is the bit length of the register. Most PLC shift register instructions can be programmed to operate the outputs from as little as 1 up to as many as 64 at a time if required.

4.3 Saving programs to disk files

If you want to save your programs on to a floppy disk, you must make sure that it is formatted first. To format a floppy disk, proceed as follows using the MS-DOS command

C\>FORMAT A:

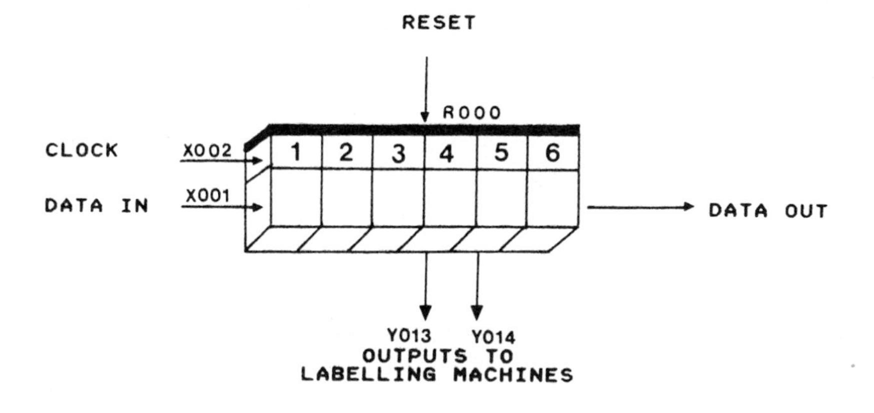

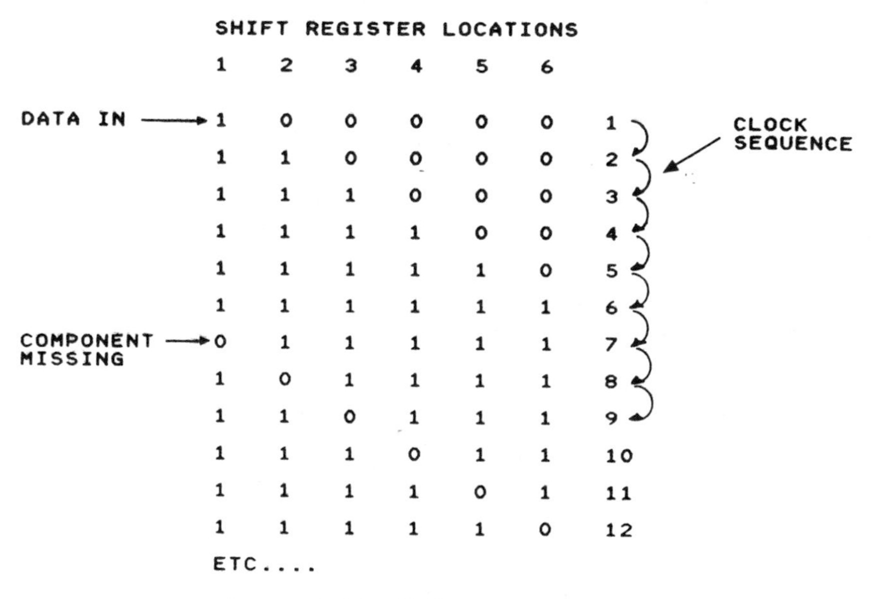

SHIFT REGISTER LOCATIONS

	1	2	3	4	5	6	CLOCK SEQUENCE
DATA IN →	1	0	0	0	0	0	1
	1	1	0	0	0	0	2
	1	1	1	0	0	0	3
	1	1	1	1	0	0	4
	1	1	1	1	1	0	5
	1	1	1	1	1	1	6
COMPONENT MISSING →	0	1	1	1	1	1	7
	1	0	1	1	1	1	8
	1	1	0	1	1	1	9
	1	1	1	0	1	1	10
	1	1	1	1	0	1	11
	1	1	1	1	1	0	12

ETC....

Figure 4.10 Shift register operation

Then, press [ENTER] and wait for instructions which will appear on the screen. If you have two drives on your computer, i.e. drive A and drive B, check which drive you are using to format your disk. On completion the computer will ask you if you want to format another disk; answer 'yes' or 'no'.

The floppy disk is now ready to use and store your programs. One word of warning: many students have formatted disks by mistake and have lost all their work, so make sure you format disks carefully.

When you are ready to save programs to your floppy disk, make sure that you change the path from the hard disk drive C:, to either drive A: or drive B: – I have known students who think they have saved their work to the floppy disk

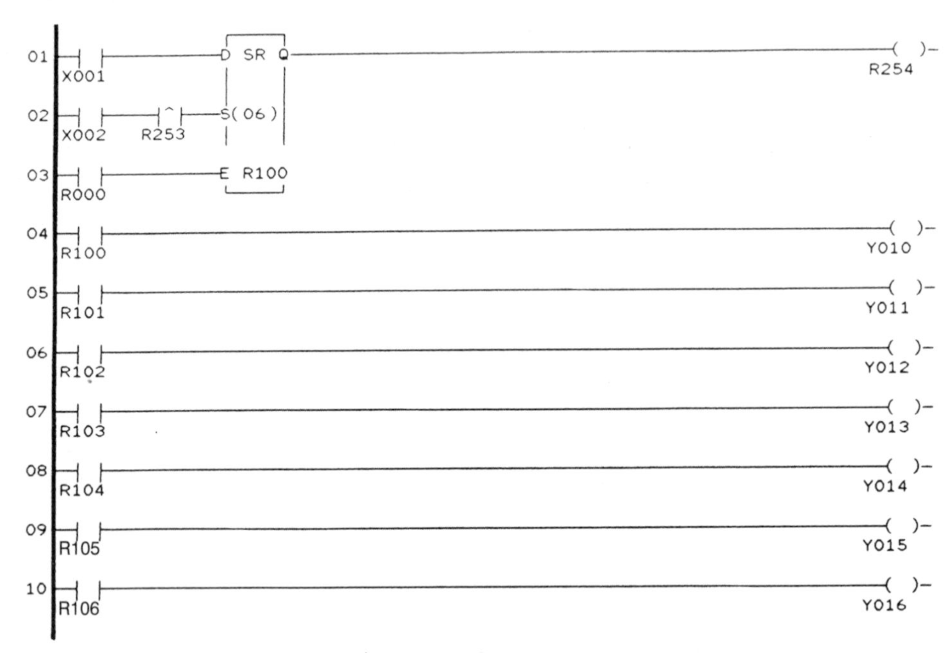

Figure 4.11 Shift register program

and a week later they cannot find the files. Refer to Appendix 3 for typical DOS commands.

4.3.1 LOAD FROM A DISK FILE TO A PLC UNIT

This will be the procedure to load a ladder logic program which is stored in a disk file and transmitted to a PLC unit on-line. This operation must be performed according to the PLC instructions you are using, i.e. the manufacturer's instructions.

4.3.2 SAVE FROM A PLC UNIT TO A DISK FILE

This function reads a ladder logic program from a PLC unit on-line and stores it in a disk file. This operation must also be operated according to manufacturer's instructions.

4.3.3 SAVING TO A PLC BATTERY-BACKED RAM

All PLCs are fitted with random access memory (RAM) and sometimes they use a battery back-up in the event of a power supply failure. If this is the case, there will be a warning lamp on the PLC unit to indicate a good or bad battery. Therefore, make sure the lamp is indicating a good battery before you proceed to save your programs to the PLC unit.

4.3.4 SAVING A PROGRAM TO EEPROM AS A BACK-UP

Some PLC units use electrically eraseable programmable read-only memory (EEPROM). This device does not need a battery back-up and can be relied upon to save ladder logic programs indefinitely when the power goes off. The memory cells can be erased electrically and modified by writing to them with a new program. Therefore, as the stored data cannot be lost, programs can easily be copied from the EEPROM. Writing to the EEPROM can only be performed with the PLC unit switched to HALT mode.

4.3.5 COMPARING PLC CONTENTS WITH MASTER PROGRAM

When you call up this function, assuming that your PLC programming software has this function, the comparison can be displayed on the screen or directed to a printer. A number of areas can be compared, for example:

- ladder logic
- I/O allocation tables
- register values.

When 'compare' is selected, the result will either be 'Compares' or 'Miscompares', for each page of the ladder logic program, with the differences displayed.

4.4 Testing and editing a PLC program

When you have entered and saved a ladder logic program to your PLC unit, the next phase is to test it with the PLC unit on-line to a computer. You will be able to observe the ladder logic program simulate the actions of the inputs and outputs.

Any discrepancies in the program can now be changed or modified, i.e. the preset time in the 'Timers', changing the normally open contacts to normally closed contacts, or a modification to the program control logical sequence.

Now when you have finished testing and modifying your ladder logic program on-line to the PLC unit, you must save the changes to your master disk files. It is quite easy to forget to perform this operation and valuable time as well as data can be lost if you fail to save to your disk files.

4.5 Forcing input and output devices

This function allows you to enable or disable a forced condition on any device. The force function allows contacts and coils to be turned ON or OFF from the computer keyboard, rather than the I/O status in the PLC program.

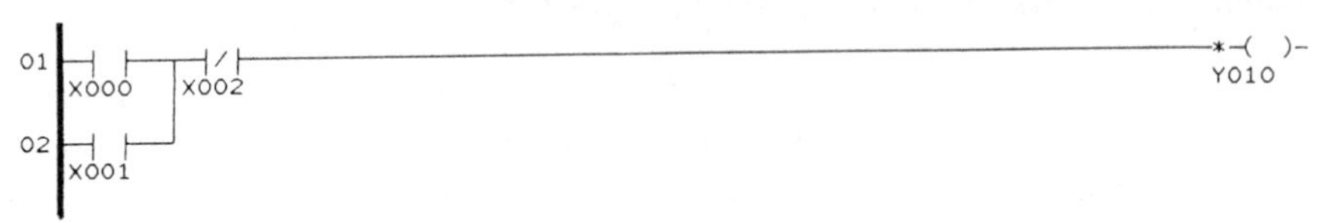

Figure 4.12 Forced coil

To force a device ON, call up the device code by using the function keys (F1–F12), i.e. 'R', 'X' or 'Y' followed by the number, and press [Enter]. The device can then be turned OFF when required in a similar way. Figure 4.12 shows an output device which has been selected as a forced option (the '*' symbol denotes a forced coil). This is particularly useful when testing PLC programs with production equipment.

Note that the procedure for operating this function will vary according to the PLC manufacturer's instructions.

4.6 Summary and test questions

This chapter has been dedicated to writing and proving ladder logic programs. The programs shown have been developed and tested using the Toshiba EX-PDD software. If you use a different software package you may find it necessary to modify these programs slightly to function correctly as described. Ladder logic is widely used and it can easily be interpreted from one software package to another.

1. What is the following table used for
   D000 C00 R000 T000
   D001 C01 R001 T001
   D002 C02 R002 T002
   D003 C03 R003 T003
   etc ...?
   (a) I/O number system
   (b) the availability of internal devices
   (c) PLC memory capacity
   (d) operational sequence.
2. What do the following symbols mean?
   (a) PLC format
   (b) I/O available

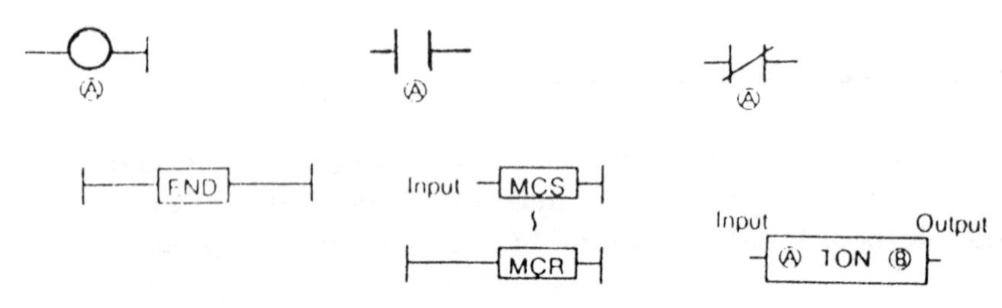

(c) PLC keycodes (icons)
(d) address location of devices.

3. Which internal relay will be ON after 10 counts?

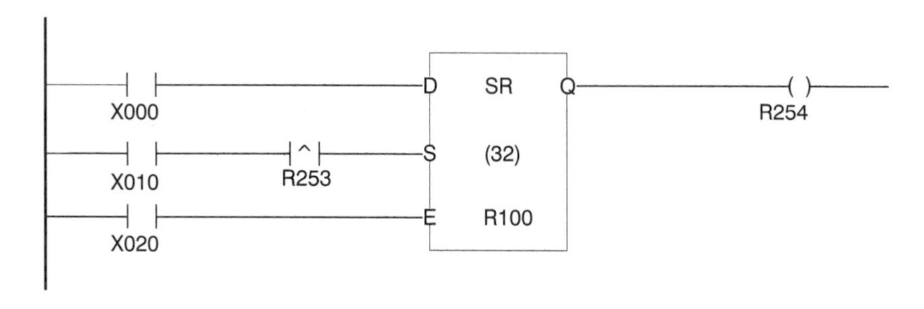

(a) R108
(b) R109
(c) R110
(d) R111.

4. What is the function of the diagram shown?

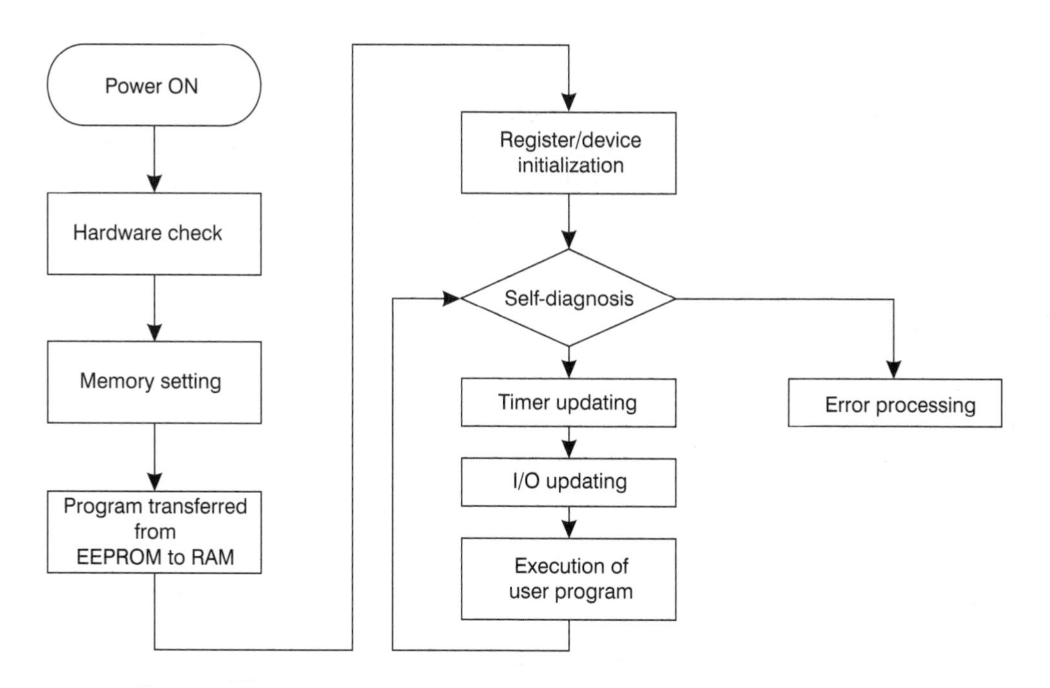

(a) logic diagram
(b) circuit diagram
(c) flow chart
(d) algorithm.

5. The timer has been changed from

   [00100 TON T000]

   to

   [00110 TON T000].

   What values were present before and after the change? The time is set in 0.1 s units.
   (a) 5 s and 10 s
   (b) 10 s and 11 s
   (c) 100 s and 110 s
   (d) 1 s and 1.1 s.
6. When a PLC system is connected on-line to a PC, which of the following is a correct statement?
   (a) Store the floppy disk in a safe place.
   (b) Enter a new program from the disk file to the PLC memory.
   (c) The PLC can control the PC.
   (d) The PLC can print out the ladder logic.
7. Write a ladder logic program for a security system consisting of a number of sensors connected to doors, windows and grills, etc. A timing device enables the operator to reset the system within a 30 s delay and an audible alarm. When one of the sensors has detected a break, the program should be activated and the timing device started to count down from the preset time. The audible alarm should go ON if the reset switch is not operated within the preset time.

5
Interfacing the PLC with Production Equipment

5.1 PLC protocols and communications

In this chapter we are going to look at ways of communicating with PLC units and how they can be connected to production equipment, i.e. transducers and sensors, etc.

From Chapter 2 and Chapter 4 you should be able to understand the basic principles of programming the PLC and how to save the programs to a floppy disk as a back-up. Communications can thus be defined as the means of transferring digital information in the form of 1's and 0's from a computer terminal (the transmitter) to a PLC unit (the receiver) and vice versa.

Take a close look at Fig. 5.1 and Fig 5.2. They are the two most common types of communication protocols associated with connecting a PLC to a computer: RS-232C and RS-422A. RS-232C is used for short distances and RS-422A is used for long-distance communications in, say, a large factory. It is important to remember which type to use, as RS-232C can only be used in close proximity to a computer terminal. Also, the RS-422A system can be used as a bus system so that the computer terminal can communicate with more than one PLC unit; see Fig. 5.3. The reasons will now be explained in more detail.

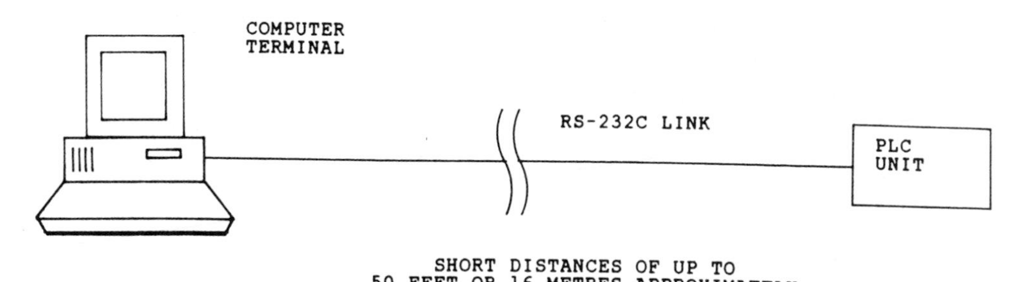

Figure 5.1 RS-232C link

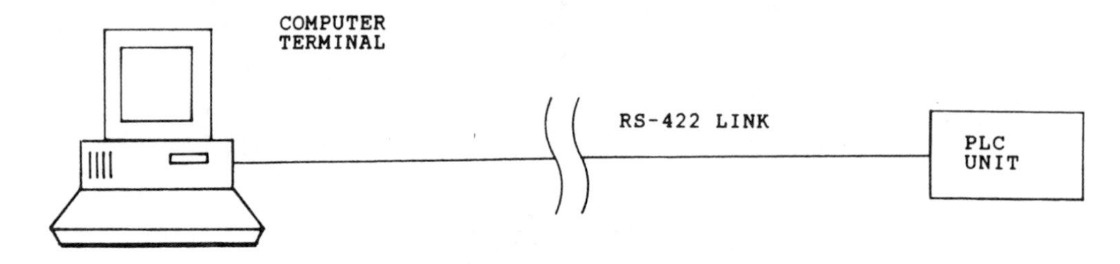

Figure 5.2 RS-422A link

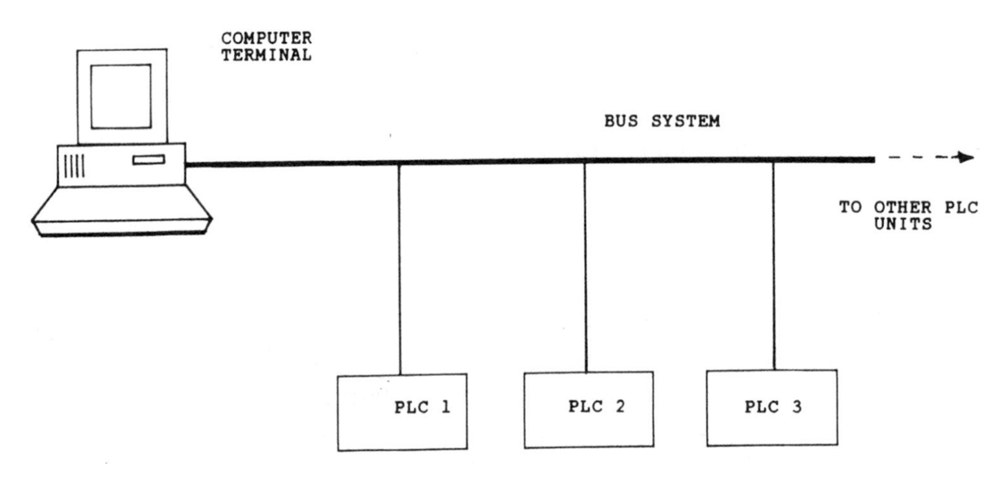

Figure 5.3 RS-422A bus system

5.1.1 RS-232C INTERFACE

This is probably the most common interface for serial transmission between a computer terminal and production equipment with devices. Figure 5.4 shows the logic signal levels and Fig. 5.5 shows the minimum line connections required to connect to a PLC. However, the most common arrangement can be seen in Fig. 5.6 using pins 1–7 and 20 of a 25-way D-type connector.

The RS-232C mode of operation is single ended, i.e. it uses an unbalanced line and will only operate with one driver and one receiver. The maximum data rate is about 20 Kbits per second.

I have previously indicated that the RS-232C interface can only be used for a short distance of up to 16 m. For greater distances, telephone lines are used with the addition of a modem at each end of the line.

5.1.2 RS-422A INTERFACE

The electrical characteristics of the RS-422A interface is of a balanced digital circuit. This means that it is isolated from the earth or common ground connec-

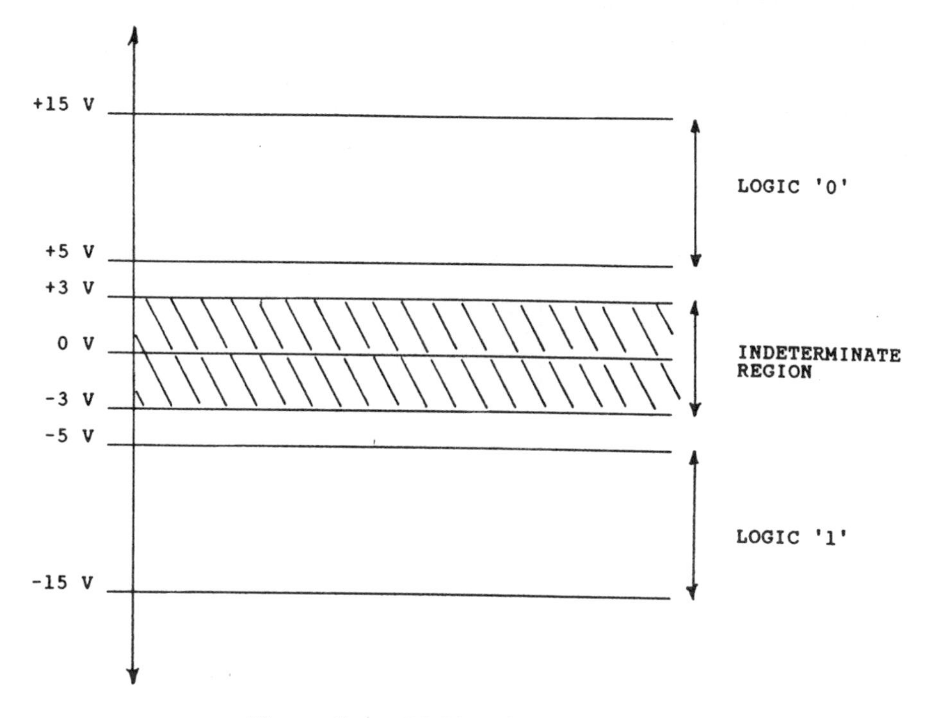

Figure 5.4 RS-232C logic signal levels

tion; see Fig. 5.7. This is an improvement on the RS-232C specifications to provide better line matching. The line uses a twisted-pair cable, i.e. two wires for each signal with a common ground and screen.

RS-422A allows for higher transmission rates as compared with RS-232C because individual balanced wires are used for each signal and, as the transmission system is balanced, ground requirements are less critical. Also, up to 10 receivers (PLCs) can be accommodated from a single driver on the computer.

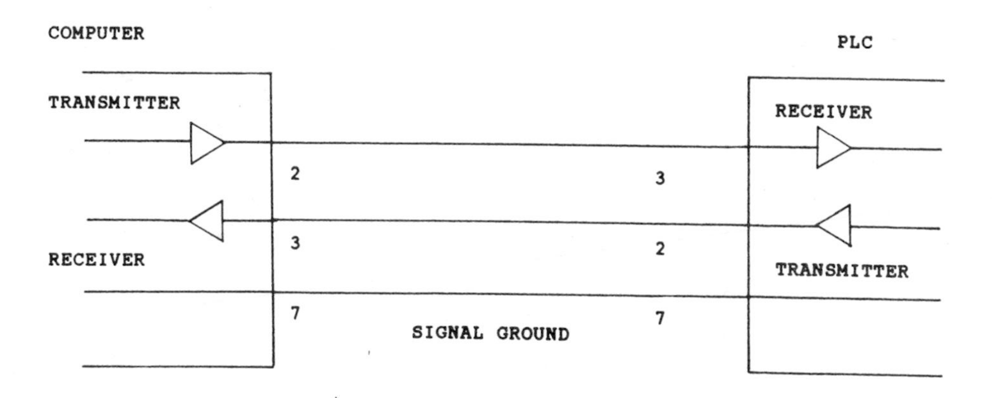

Figure 5.5 RS-232C showing minimum line connections

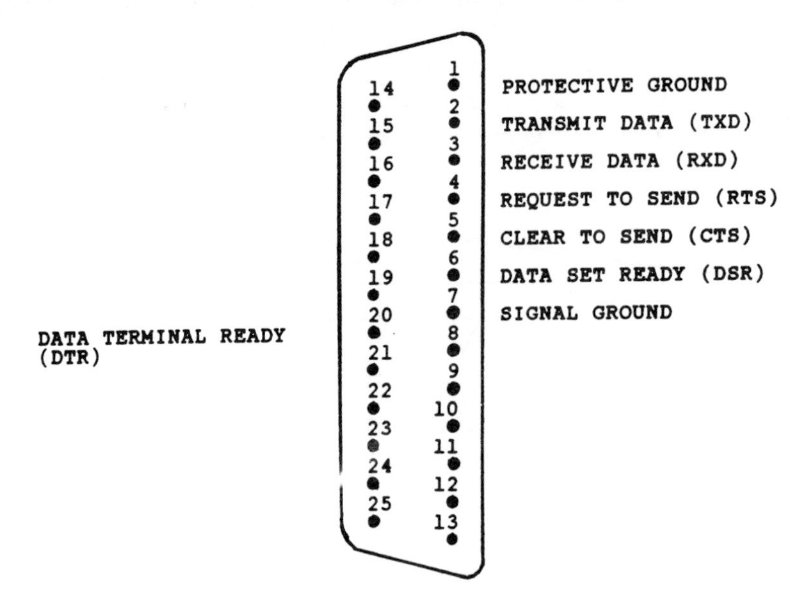

Figure 5.6 RS-232C pin connection

Sometimes, due to the non-standard markings on production equipment, you may need to swap the A and B wires to get the link to operate.

5.1.3 IEEE-488/GPIB BUS INTERFACE

The IEEE-488 (Institution of Electrical and Electronic Engineers) bus was originally developed by Hewlett-Packard and is currently known as the General Purpose Instrument Bus (GPIB).

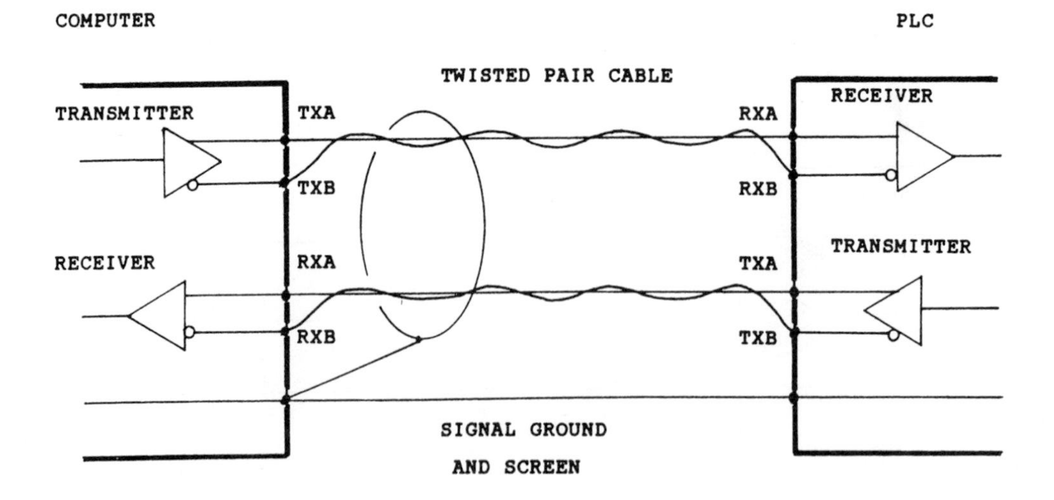

Figure 5.7 RS-422A balanced line connections

This system was developed to connect a range of instruments in automatic test equipment (ATE) and where data can be exchanged between a number of devices; see Fig. 5.8. It consists of a number of control lines and a data bus. The computer is the controller which provides processing facilities as well as the management of data on to the bus.

It is also important to realize that this bus system can provide for the data transfer of different devices having different response times; therefore it is the slowest device which will determine the rate at which data transfer takes place. However, only one talker (the PLC transmitting) can be active at a given time, but it is possible for several listeners, i.e. the PLCs, to be receiving data all at the same time from the computer.

5.1.4 TWISTED-PAIR CABLE

Twisted-pair cable is made from strands of wire twisted in pairs; see Fig. 5.9. It is extensively used for telephone and cable communications and each twisted pair can carry a single telephone call between two people as well as two machines.

It can be used for analogue as well as digital transmission. For long-distance lines the use of a number of repeaters will refresh the signal to maintain data integrity.

5.1.5 COAXIAL CABLE

Coaxial cable is resistant to transmission interference which can corrupt data, as it provides a fast, relatively interference free transmission medium; see Fig. 5.10. Its construction consists of a central conductor core which is surrounded by a layer of insulating material. The insulating layer is then covered by a conducting shield or screen, with a further insulating layer to protect it.

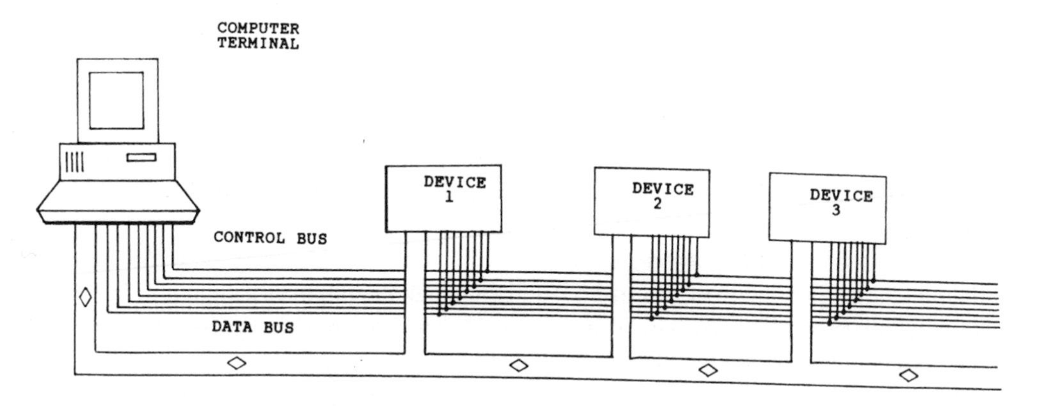

Figure 5.8 IEEE-488/GBIP bus system

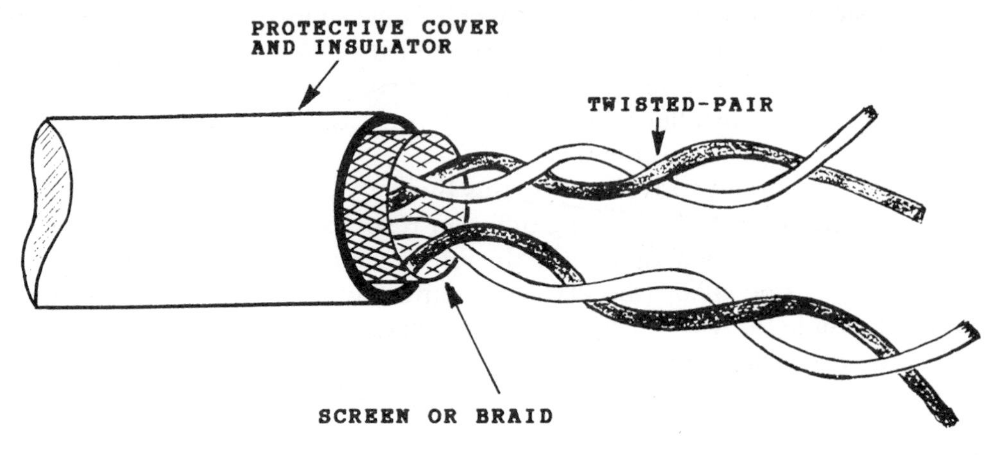

Figure 5.9 Twisted-pair cable

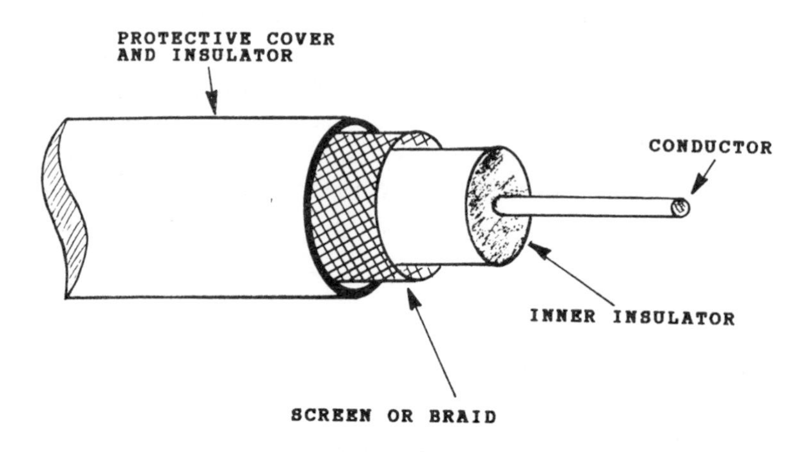

Figure 5.10 Coaxial cable

There are many different types of coaxial cable available and they vary in quality from being very flexible to being rigid for long-range communications.

5.1.6 OPTICAL FIBRE CABLE

This consists of many clear glass fibre strands which can transmit light or infra-red rays instead of electrical signals. The signal or data is transmitted by a light-emitting diode which has two states: a logic '0' = OFF and a logic '1' = ON. The construction of the fibre optic cable makes it possible for the transmission of millions of bits per second. Figure 5.11 shows the construction; each strand can carry different information simultaneously. At the receiving end the light is converted back into an electrical signal by a detector.

The advantage of optical fibre links is that they are impervious to all types of electrical interference and they also provide electrical isolation between the

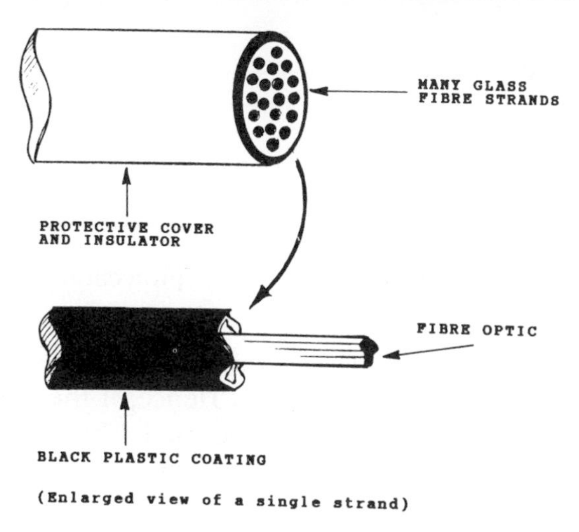

Figure 5.11 Optical fibre cable

sender and the receiver, preventing transient currents reaching the microprocessor circuits. Also, they have a very high frequency characteristic, typically of the order of 300,000 GHz, which permits much more information to be transmitted over a single channel.

Optical fibre cables are currently more expensive than twisted-pair and coaxial cables, but they are very popular for long-distance communications where security and integrity of data is required.

5.2 Interfacing techniques

The input signals that are sensed by a PLC system are generally derived from mechanical contacts and switches. These contacts and switches will only signal an open circuit or a closed circuit. Therefore, the input interface to the PLC will require an external voltage supply to the input driver. Similarly, an external voltage supply to the output interface will also be required.

Interfacing techniques can take many forms, depending on the application of the field device that is to be connected to the PLC. The remainder of this chapter will be dedicated to the interfacing of field devices to the PLC unit.

Note that some of these interfaces are usually part of the PLC circuitry, but in some cases it may be external. Therefore, it is a good idea to check the manufacturer's data and specifications for that particular PLC system.

5.2.1 PROTECTION METHODS

Figure 5.12 shows a high-voltage protection circuit to guard against a very high voltage. The zener diode is used to clamp the input voltage to a safe level. The fuse type is quick-blow as in the event of a very high voltage appearing on the input the zener diode will conduct more current thus blowing the fuse.

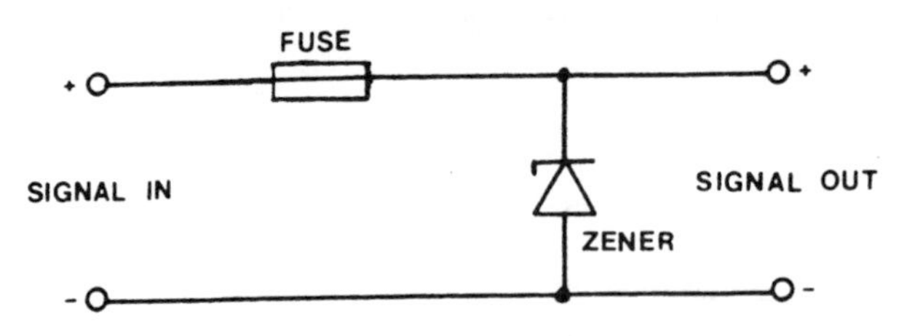

Figure 5.12 High-voltage protection

Figure 5.13 is a reverse polarity protection circuit. This can be very useful when a.c. supplies are used as well as d.c. Hence, if the inputs to a PLC are described as a.c./d.c., then this circuit will be used. The diode shown will only conduct one way, i.e. in the direction of the arrow.

In Fig. 5.14 one method is shown of how to protect the semiconductor switching device in the output stage. D1 is for reverse polarity protection and is usually part of the PLC circuitry. D2 is always connected across the relay coil externally. This is general practice in order to remove the transient current that is generated by the relay coil when the power to it is switched off. The transient current is known as induced back-emf.

5.2.2 OPTO-ISOLATORS

These are primarily used to isolate between a switch network and the input logic circuits in a PLC unit. A basic layout is shown in Fig. 5.15.

The reasons for using opto-isolators are as follows:

- to isolate the low-voltage input circuits from the switch contacts in a much larger supply, e.g. 230 V a.c. mains supply;
- to provide electrical isolation between the switch input signals in a multi-input system;
- to isolate the external wiring and components from the digital circuitry and minimize the possibility of damage to the PLC system;
- to eliminate ground-loop and earth-fault problems which can cause noise and interference.

A typical arrangement of a single opto-isolator suited for fast signal transfer and offering excellent noise immunity is shown in Fig. 5.16. This circuit, consisting

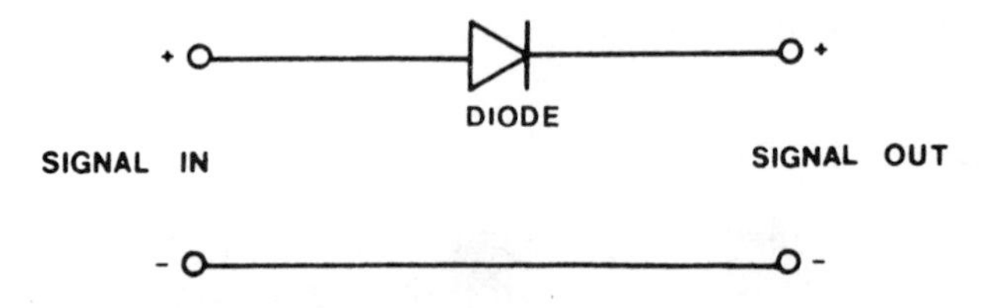

Figure 5.13 Reverse polarity protection

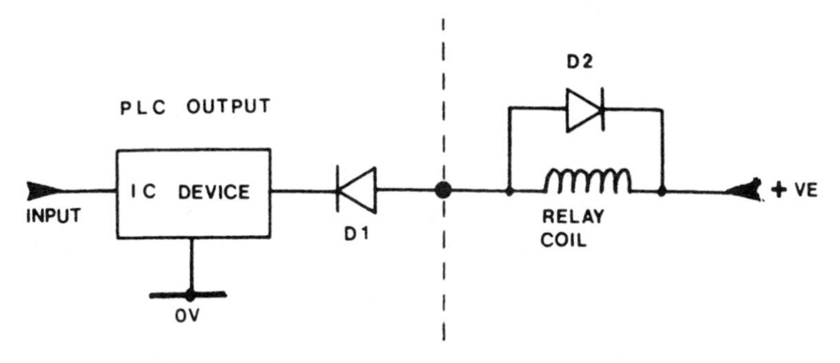

Figure 5.14 Protection on output switching

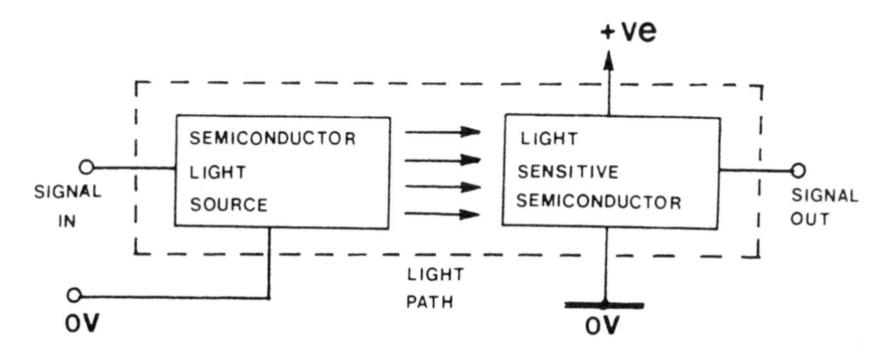

Figure 5.15 Basic layout of opto-isolator

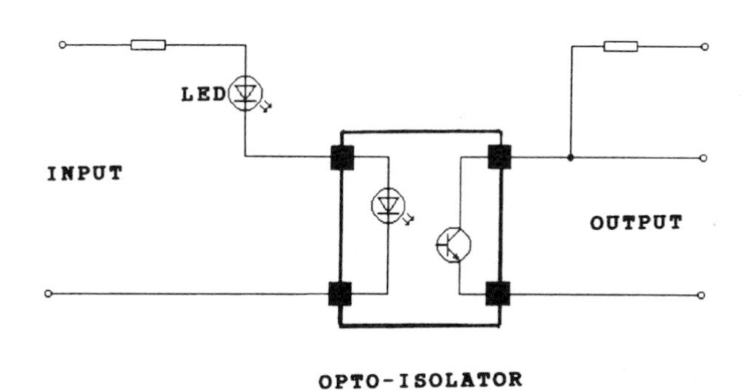

Figure 5.16 Single opto-isolator circuit

of a light-sensitive transistor and a light-emitting diode, can be configured for the output as well as the input interface of a PLC unit, thus allowing data to be transferred between the two electrically isolated systems. In addition, opto-isolators come in different packages to suit many applications, e.g. dual opto-isolators and a quad opto-isolators.

The external LED (light-emitting diode) is to give you an indication of the status of the input and/or the output, i.e. the LED on means that the signal is present.

5.2.3 SWITCHING RELAYS AND SOLENOIDS

In many cases the output stages of a PLC use a solid state semiconductor device, e.g. a transistor, a thyristor or a triac. These devices are designed to switch large currents that are sometimes required of both the relays and the solenoids.

However, care must be taken to connect the relay or the solenoid to the PLC in the correct mode, i.e. SINK or SOURCE. Figure 5.17 shows the two modes of connecting a relay to the output terminal of a PLC. In the transistor sink mode the +24 V supply is made the common connection, whereas in the transistor source mode the 0 V line is common.

Again, I must emphasize the fact that you should always check the manufacturer's specifications to see which mode the PLC output is configured to, prior to connecting it up to production equipment.

5.2.4 SEMICONDUCTOR SOLID STATE RELAYS

With the need for better switching and demands for electrical isolation between production equipment and the PLC, the solid state relay has evolved to become the most popular device since the electromagnetic relay. It has no moving parts or mechanical properties that can wear out during its useful life. The device is totally electronic and it uses a range of semiconductors and components to accommodate switching and isolation between the PLC and production equipment.

Figure 5.18 shows a solid state relay device. It can be connected to the output of a PLC terminal in the sink mode and also in the source mode. Note the device connections.

The solid state relay consists of an opto-isolator, a full-wave bridge rectifier and a triac switch. The capacitor and resistor across the triac is a suppressor, in order to remove the transient current when the triac switches the load on and off. Figure 5.19 shows a typical circuit diagram for this solid state relay device. The device is totally enclosed or encapsulated. They can take many forms as there is no standard physical size. The switching power is available from as little as 1 Ampere up to 100 Amperes.

Note that the solid state relay will require a heat sink to dissipate the heat energy when it is operating at very large currents and the manufacturers of these devices will always recommend the correct type you should use, together with the fixing details and suitable ventilation.

5.3 Photo-electric sensors

These devices are very popular in factory automation, their main purpose being to detect the presence of components moving along on a conveyor belt. They

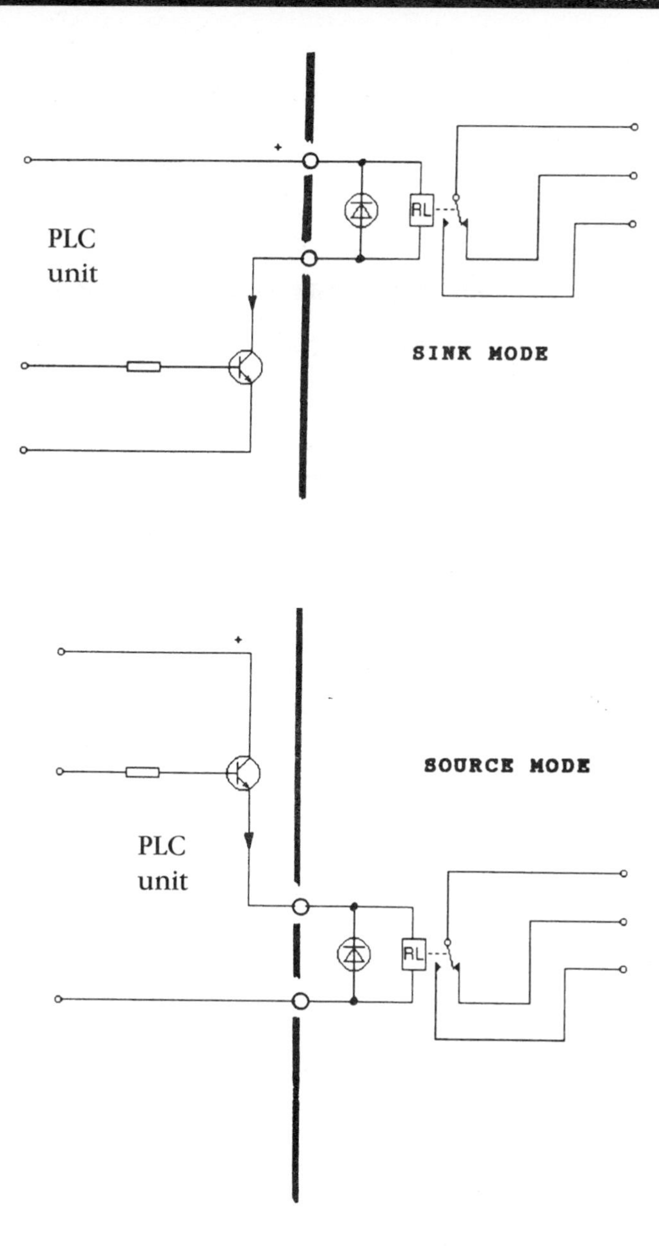

Figure 5.17 Relay connections to a PLC

can be used to count the number of components flowing through the manufacturing process and to detect whether a component is correctly assembled. They can initiate the start of another process on the assembly line when required; for example, a labelling machine.

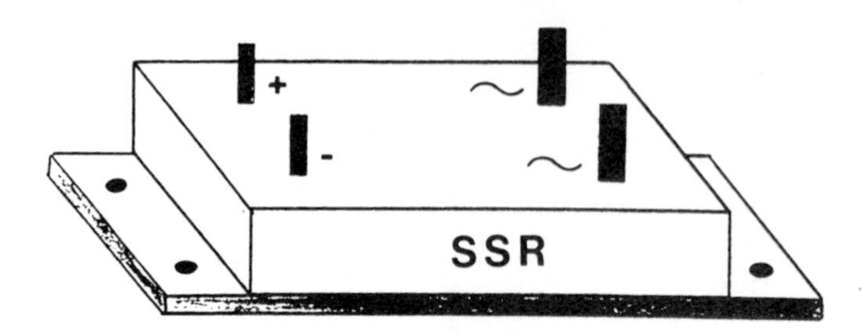

Figure 5.18 Solid state relay device

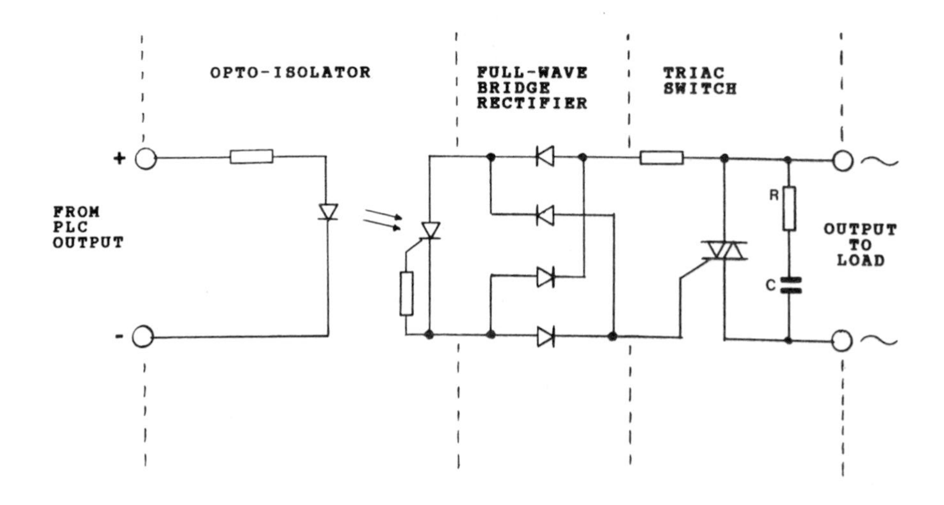

Figure 5.19 Typical solid state circuit

There are two ways of detecting components moving along the production line:

- detection by blocking the light from the transmitter to the receiver;
- detection by reflecting the light from the transmitter to the receiver.

These can be further categorized as:

- through-beam system (blocking)
- reflex system (blocking)
- diffuse system (reflection).

See Figure 5.20.

The through-beam and reflex types must be positioned carefully, because of the tolerance of the receiver and the reflector in each case. A component passing through the light beam will break the circuit and change the output from the sensor.

THRU-BEAM SYSTEM

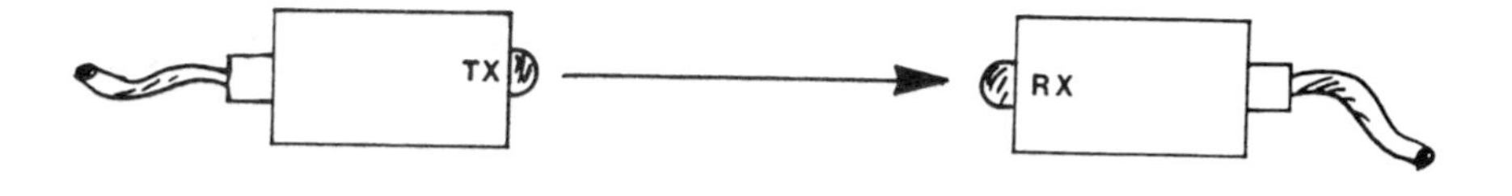

REFLEX SYSTEM

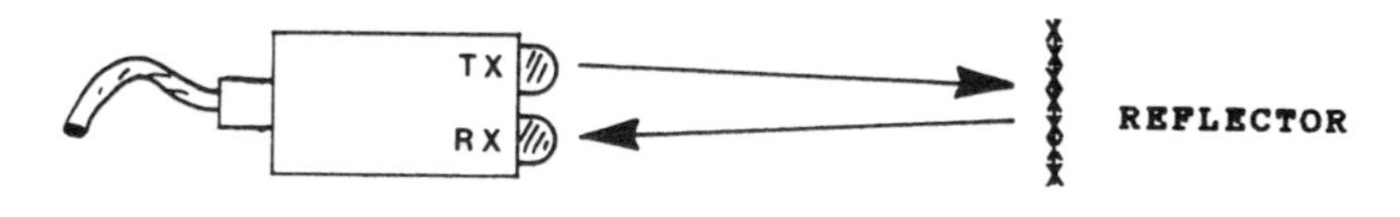

DIFFUSE SYSTEM

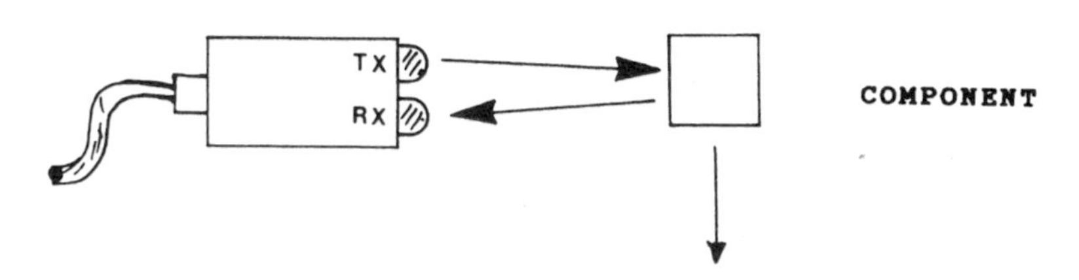

NOTE: TX = TRANSMITTER
RX = RECEIVER

Figure 5.20 Photo-electric sensor operations

With the diffuse type, however, any reflecting object within close proximity to it will be detected and the sensitivity of the sensor can be adjusted so that the background is not detected.

The set-up and testing of these photo-electric sensors is made quite simple: you just choose the correct type of sensor and reflector for each application. Also, they do not respond to normal ambient light in the factory because they use infra-red light-emitting diodes and infra-red semiconductor sensors.

Figure 5.21 shows an NPN sensor connected to the PLC input. Note that because the sensor is NPN it can only be connected with the PLC configuration shown. You will notice that the sensor output is in the transistor SINK mode.

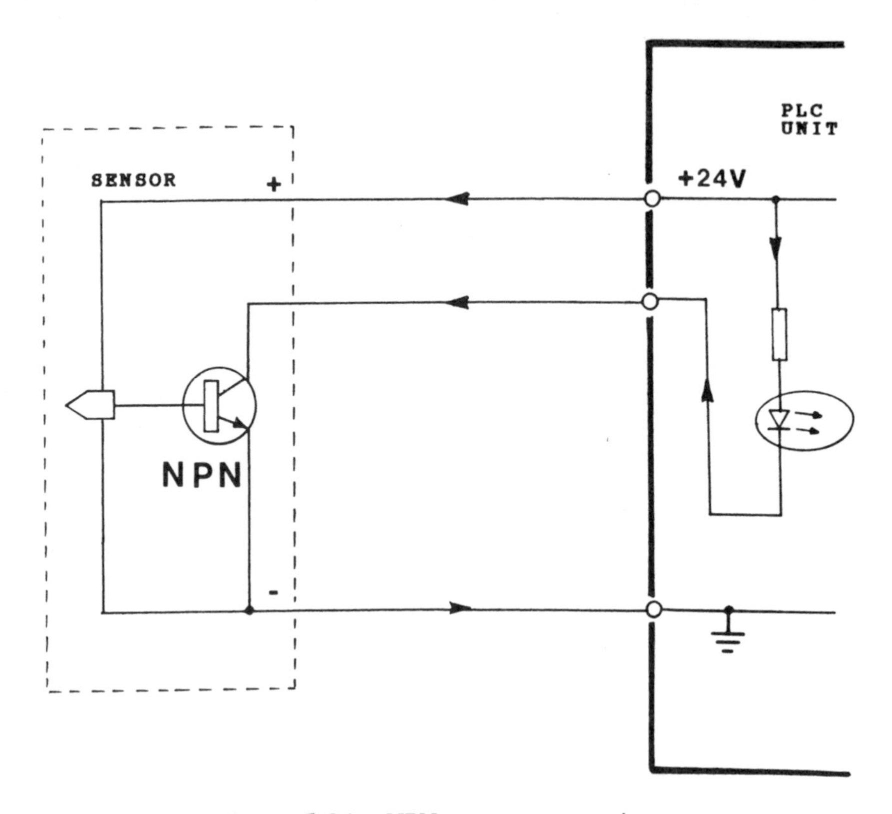

Figure 5.21 NPN sensor connections

For connection to a PLC with a PNP sensor, see Fig. 5.22, notice the difference: the transistor is now in the source mode.

For both circuits the PLC employs an opto-isolatator and I must emphasize again the importance of checking the manufacturer's specifications in choosing the correct type of sensor.

If dry output contacts are required then you will use a relay output as shown in Fig. 5.23. Notice that you have a choice of either normally open or normally closed contacts. This configuration is also useful for change-over, i.e. switch one input on and another input off.

5.3.1 OTHER PROXIMITY SENSORS

These devices are either inductive or capacitive and will detect the presence of components and other objects in a similar way to the photo-electric sensor.

Also, their shape and size is similar to the photo-electric sensor and it is important that you check the specifications of the sensor before you either use it or replace it.

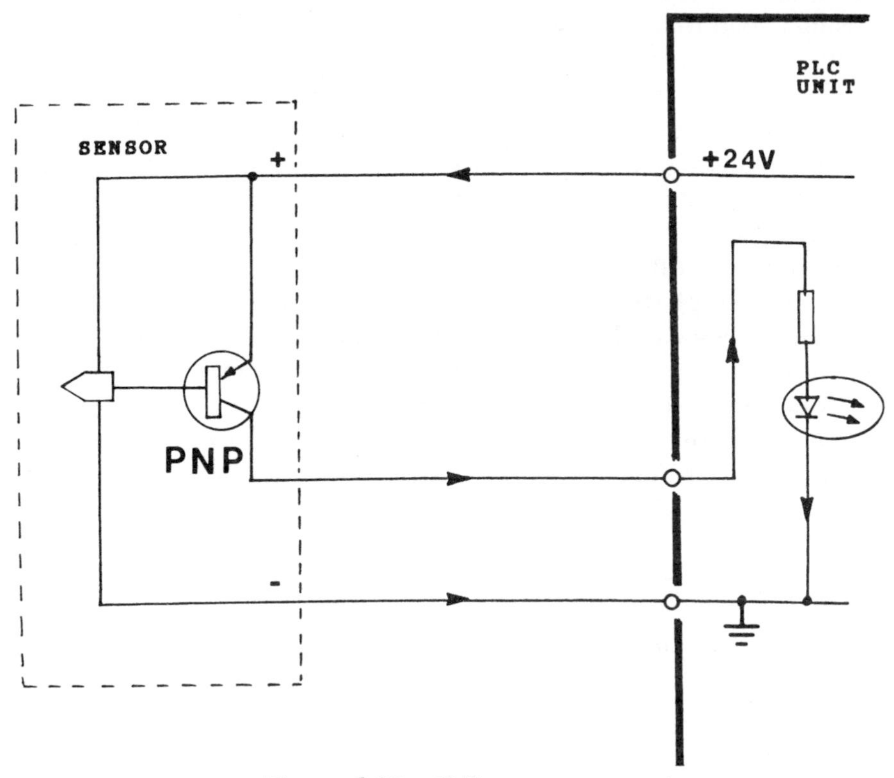

Figure 5.22 PNP sensor connections

The inductive proximity sensor operates on the emission of a radio-frequency electromagnetic sensing field from an oscillator and coil. The coil is wound on a ferrite core to concentrate the sensing field in the axial direction through the active face of the proximity sensor.

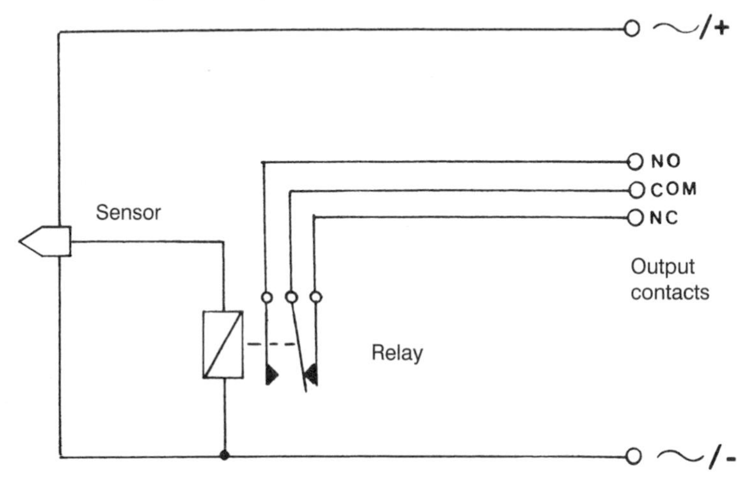

Figure 5.23 Sensor with relay output contacts

When a metallic object enters the electromagnetic field it absorbs energy from it causing a load to be placed on the oscillator, which in turn reduces the output. As the metallic object moves closer to the face of the sensor there will be further absorption of the electromagnetic field and thus a further reduction on the output. At this point the output will be low enough for the trigger circuit to switch the device output to another state. For example, if the sensor has a normally open contact it will close when a metallic object is sensed.

The capacitive proximity sensor works on the principle of using a high-frequency oscillator with a floating electrode on the face of the sensor. When a medium such as glass, plastic or wood appears within the active area of the sensor, the oscillation begins. The frequency from the oscillator is then rectified and the resulting d.c. signal is used to switch the output stage in a similar manner to the inductive sensor.

The switching sensitivity can usually be adjusted in both cases by the setting of a potentiometer providing a presettable response of the proximity sensor.

5.4 Summary and test questions

From this chapter you will have learned how to connect most sensors and output devices to a PLC unit. In the next chapter, I will concentrate more on the analogue signals that are used, as well as the actuators and different types of motors. Try and answer the following questions before you move on.

1. What are the main differences between RS-232C and RS-422A links?
2. Explain the term 'communication protocol'.
3. What are the benefits of using a bus system?
4. Why are opto-isolators used in PLC equipment?
5. Explain why a diode must be connected across a relay coil.
6. With the aid of a sketch, explain the meaning of sink and source mode.

6
Interfacing Analogue and Special-Purpose Transducers

6.1 The basic control system

The information that goes into and out of most computer systems is usually analogue in nature, i.e. the measurement and control of temperature, pressure and flow, etc. It must be understood that PLCs and computers can only deal with digital information, i.e. binary code, as previously mentioned in Chapter 3. If it is to be used in the real-world, i.e. to process analogue information and signals, it must first convert the input analogue signal into a digital format.

When the PLC or computer has processed and worked on the converted digital information, it must convert the digital signal back to analogue again, to deal with the real-world. See Fig. 6.1 for the basic control system.

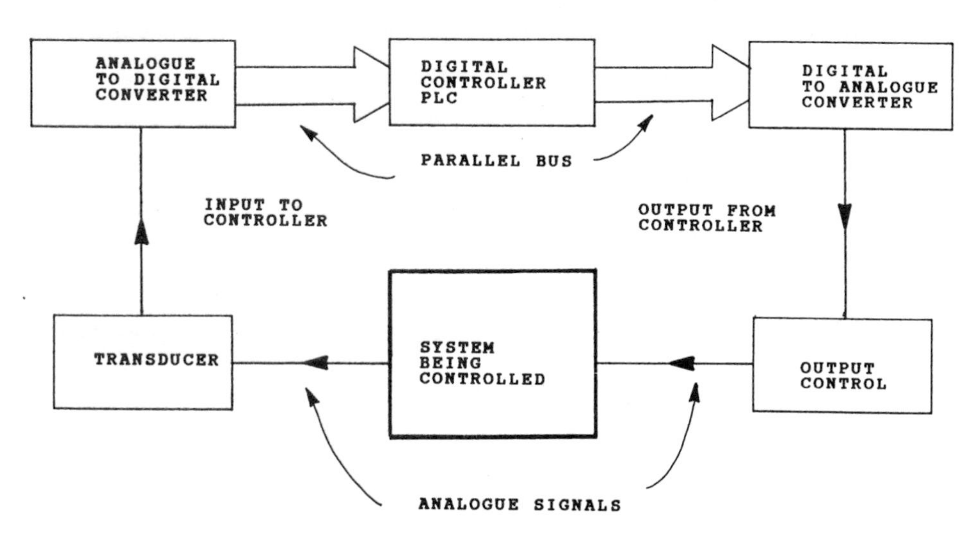

Figure 6.1 Basic control system

Note that these converters are usually separate modules and most PLC manufacturers supply them as plug-in devices as and when they are required.

6.1.1 DIGITAL-TO-ANALOGUE CONVERTER (DAC)

The digital-to-analogue converter (DAC) converts the digital signal to an analogue value. The input to the DAC is a parallel digital signal applied to a binary weighted resistor network. The weighting is proportional to the digital signal and thus the amplifier output is proportional to the digital input signal. See Fig. 6.2.

6.1.2 ANALOGUE-TO-DIGITAL CONVERTER (ADC)

The basic circuit shown in Fig. 6.3 is known as a servo or counter type. You should be able to pick out the main components of this ADC and you will also notice that the circuit is more complex than the DAC. During the conversion, the input signal is compared with a feedback signal derived from the digital output of the DAC. If an error exists, a command is issued to clock the digital counter to reduce the error to zero. This digital count value is converted by the DAC to form the analogue feedback signal. When the feedback and the input signals are equal, the counter is halted and the digital output will correctly represent the input analogue value.

The digital output is available at all times. However, while the input is changing its condition the output will be in error until the feedback is equal to the input signal. It is the PLC program that instructs the converter to begin conversion; it then waits for a signal from the converter to indicate that the conversion is complete. If the PLC reads a value before or during the conversion, then the outcome will be an incorrect signal.

I have only shown four bits for both the ADC and the DAC as an example. The resolution of a four-bit converter will only be

$$1/2^4 = 1/16 \times 100 = 6\%.$$

However, an eight-bit converter is more common and you can also program the PLC to accept any other bit-size word. A converter with an eight-bit word con-

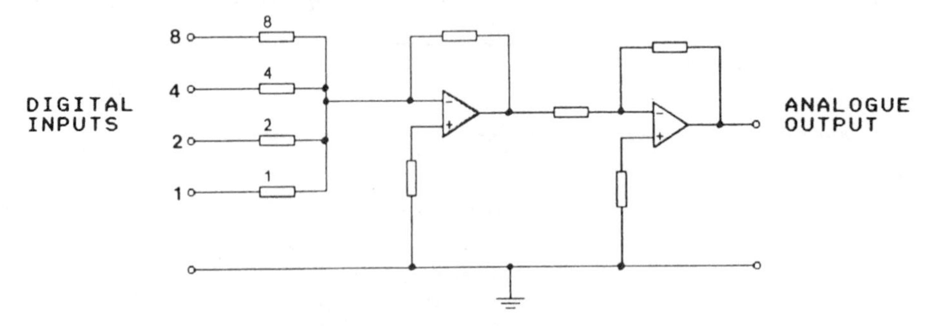

Figure 6.2 Digital-to-analogue converter

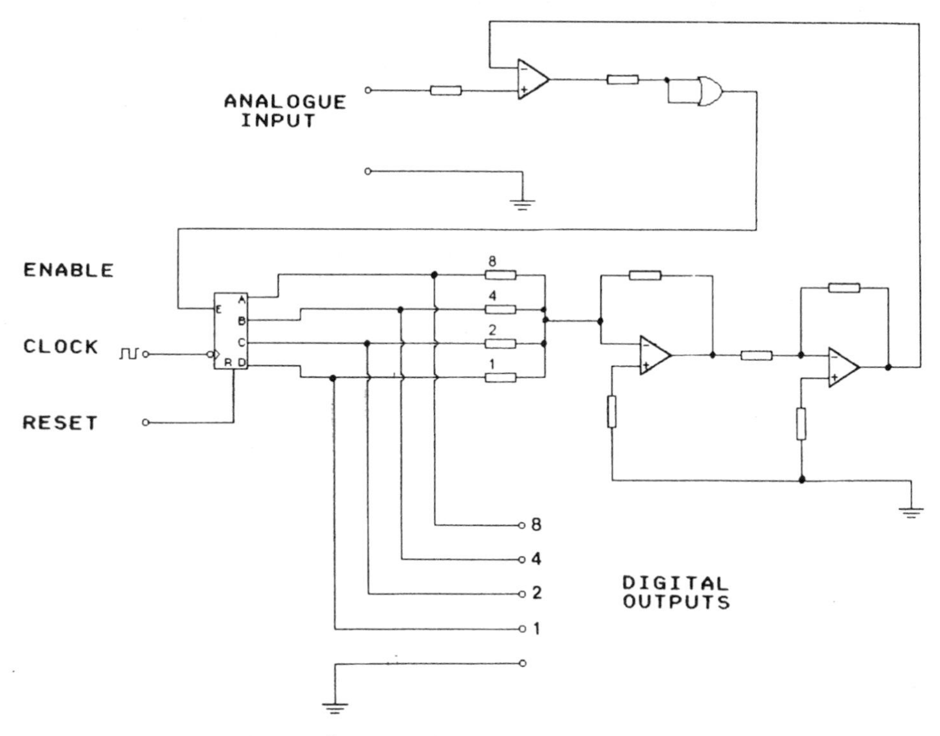

Figure 6.3 Analogue-to-direct converter

verts in eight steps, irrespective of the size of the signal and thus its resolution will be better than 0.5%. Thus

$$1/2^8 = 1/256 \times 100 = 0.39\%.$$

The DAC and the ADC will be required for most applications in this chapter, e.g. the thermocouple, the strain gauge and the servo valve, etc.

6.1.3 CURRENT LOOP TRANSMISSION

For most analogue transmission circuits the analogue signal will take the form of either a voltage or a current. For example, '0–10 V' and '4–20 mA', the latter being the most popular in industrial control systems. However, it is usual for both the input modules and the output modules of a PLC to have the ability to be connected as either a voltage or a current configuration. This makes the connection of any sensor or transducer to the PLC unit possible with ease.

In most processes where analogue signals are transmitted, a current of 4–20 mA is generally used. The offset represented by 4 mA provides you with the distinction between a zero and an open circuit, i.e. no current flow will indicate a fault. Also, the current loop can permit the connection of a few specified devices to be connected in series; see Fig. 6.4. This shows you an example of a

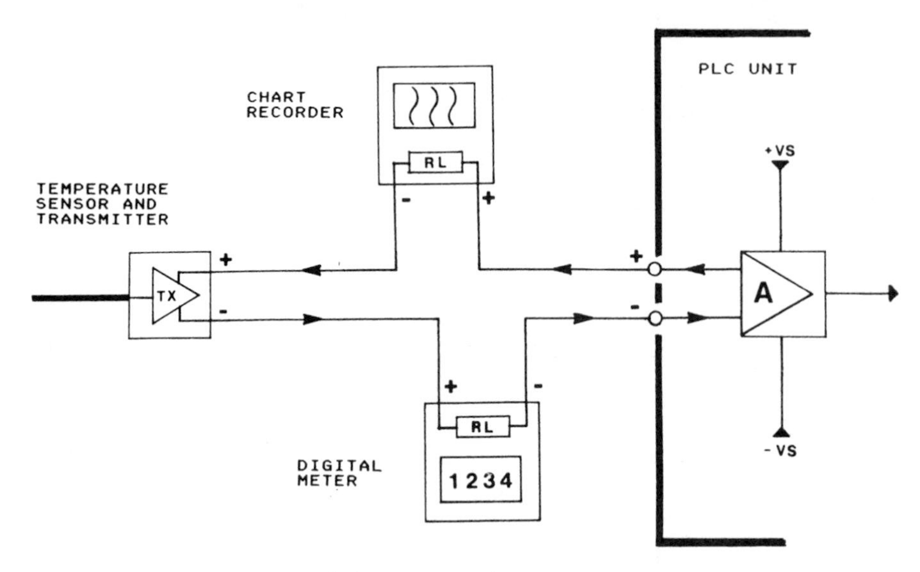

Figure 6.4 Current loop transmission

temperature sensor in series with a chart recorder and a digital meter as well as providing an input to the PLC unit.

You will also notice that the PLC controller provides the current supply to the sensor and there is no need for an external power source. The power is transmitted in one direction and the signal in the return direction of the twisted-pair cable. The devices connected on to the current loop will each have a load resistance (RL) of approximately 200 ohms.

6.2 Thermocouple

Thermocouples are made from two dissimilar metals. They are robust, economical and have good long-term stability. They are also very good at responding to changes in temperature, i.e. they have a fast response time. The principle operation of a thermocouple is to generate a voltage output which represents the temperature that it is measuring at the junction. This voltage is in millivolts (mV) and Table 6.1 gives you a comparison of some types in common use.

Table 6.1

Thermocouple type	Temperature range	Voltage range (mV)
Copper–constantan	-180 to +400	26
Chromel–constantan	0 to +980	75
Iron–constantan	-180 to +760	50
Chromel–alumel	-180 to +1260	56
Platinum–rhodium	0 to +1590	16

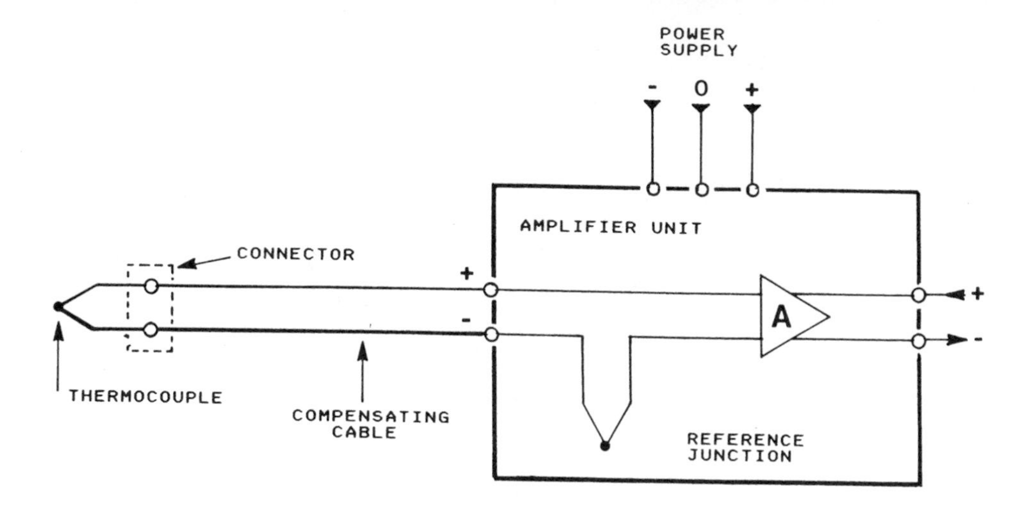

Figure 6.5 Thermocouple conditioning amplifier

You will have noticed that the output voltage is very low and therefore the connection to the input of a PLC unit must be done using a signal conditioning unit; see Fig. 6.5. It consists of the thermocouple, a compensating cable, a reference junction and an amplifier.

Compensating cable must always be used with thermocouples. For example, if you decide to use a thermocouple made from copper–constantan then you must use a cable made with the same materials in order to provide a good match. Using ordinary copper cable will result in a mismatch and cause errors in the measurement of temperature.

The reference junction is to compensate for the temperature variations at the amplifier unit. The reference point will generate a voltage and this voltage can be added to that of the thermocouple by the signal conditioning amplifier.

The amplifier will provide the necessary isolation and conversion from the small mV signal to a 4–20 mA current loop signal as described previously in this chapter.

6.3 Strain gauge

The strain gauge is mainly used for the measurement of pressure and force. Its construction consists of a metal foil gauge (in the shape of a grid) and an insulating carrier; see Fig. 6.6. The grid, or element as it is sometimes called, is resistive and its resistance will change as a force is applied to the unit to which it is attached. A tensile or compressive force can then be converted into an electrical signal.

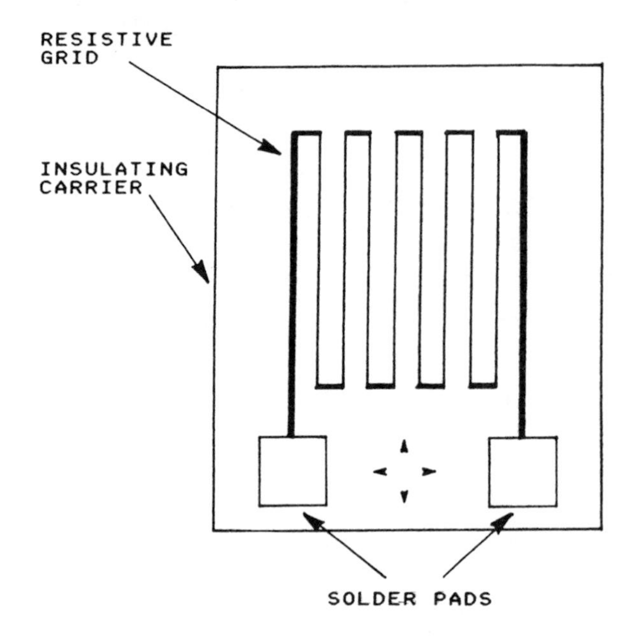

Figure 6.6 Strain gauge element

Strain gauges also need some form of signal conditioning and generally use a transmitting signal of 4–20 mA as previously mentioned for the thermocouple.

The strain gauge bridge in Fig. 6.7 is part of the sensor's circuit and this is the basic layout for pressure and force measurement. This bridge circuit is also known as a Wheatstone bridge. It is excited with a constant voltage source and the output is taken from another limb of the bridge. When a force is applied to the sensor, pressure is transmitted through the device to the strain gauge which causes an imbalance of the circuit and an output voltage proportional to the pressure is obtained.

6.4 Position encoder

Position encoders are employed wherever motion sequences are to be controlled, for example, machine tools, positioning of components, rotation and limits, and many other positioning tasks.

PLC manufacturers will usually provide a special interface module to allow position encoders to be connected as standard. Depending on the type of PLC, a single input module will consist of the following signals, 0 V, +5 V, A+, A–, B+, B–, M+ and M–.

The encoder index signal M is conditioned to give a unique and repeatable position for home. This index signal must be high when the A signal is high for the index signal M to be recognized.

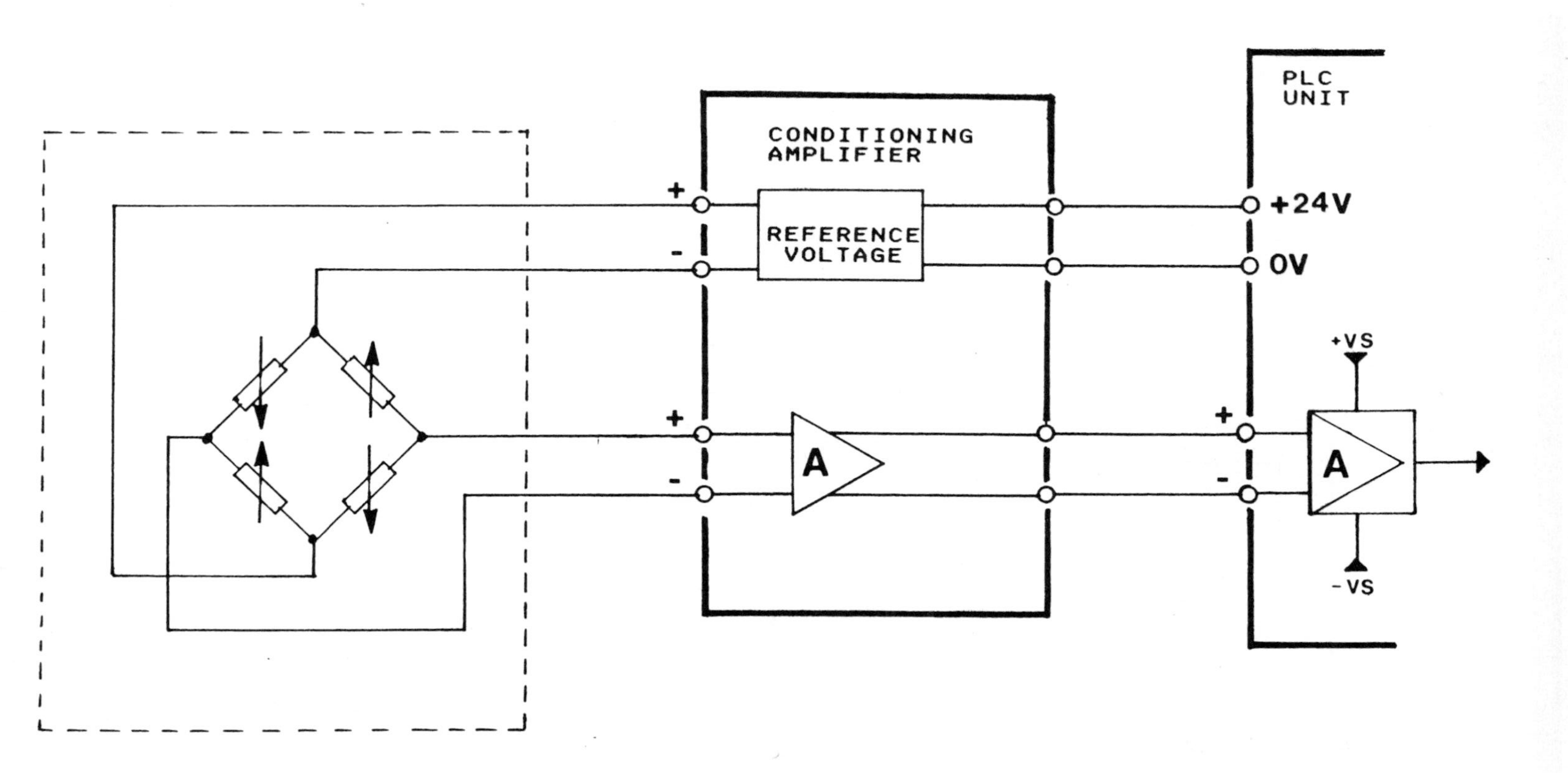

Figure 6.7 Strain gauge bridge circuit

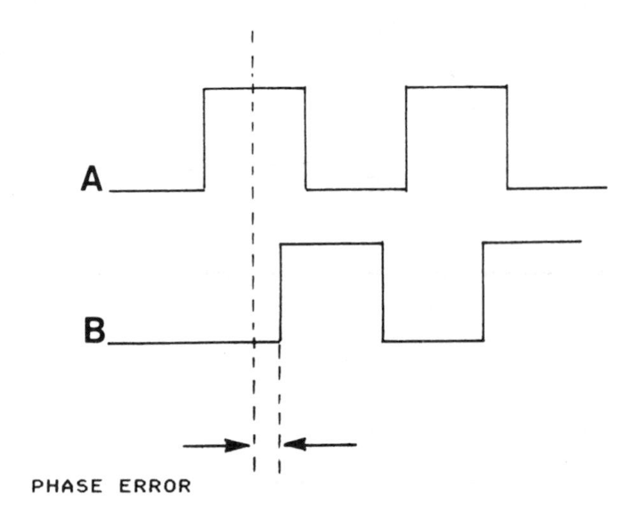

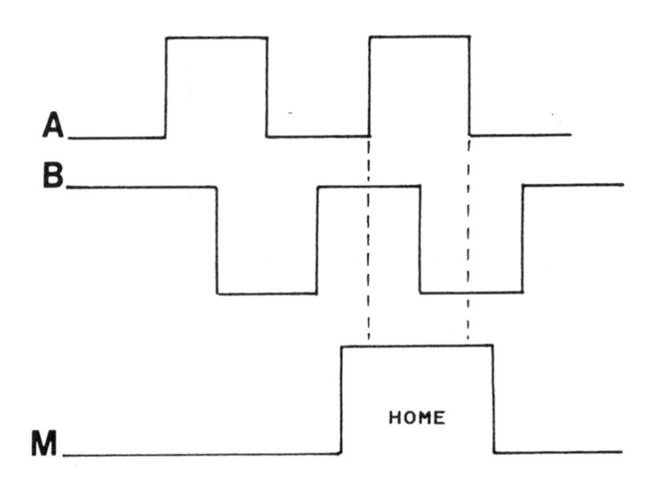

Figure 6.8 Encoder signals

The operation of an encoder can best be explained by the encoder signal waveforms shown in Fig. 6.8. It can be seen that the encoder signal A leads B for forward motion.

Figure 6.9 shows an example of connecting a position encoder to a typical PLC unit. The EXT connections are used to reset the count value to zero. You will also notice that shielded twisted-pair cable must be used with the shield or screen connected to ground.

6.5 Servo valves

Let us suppose that we need a control valve for the feedwater to the boiler of a power station. Figure 6.10 illustrates the principle of operation of the valve. In

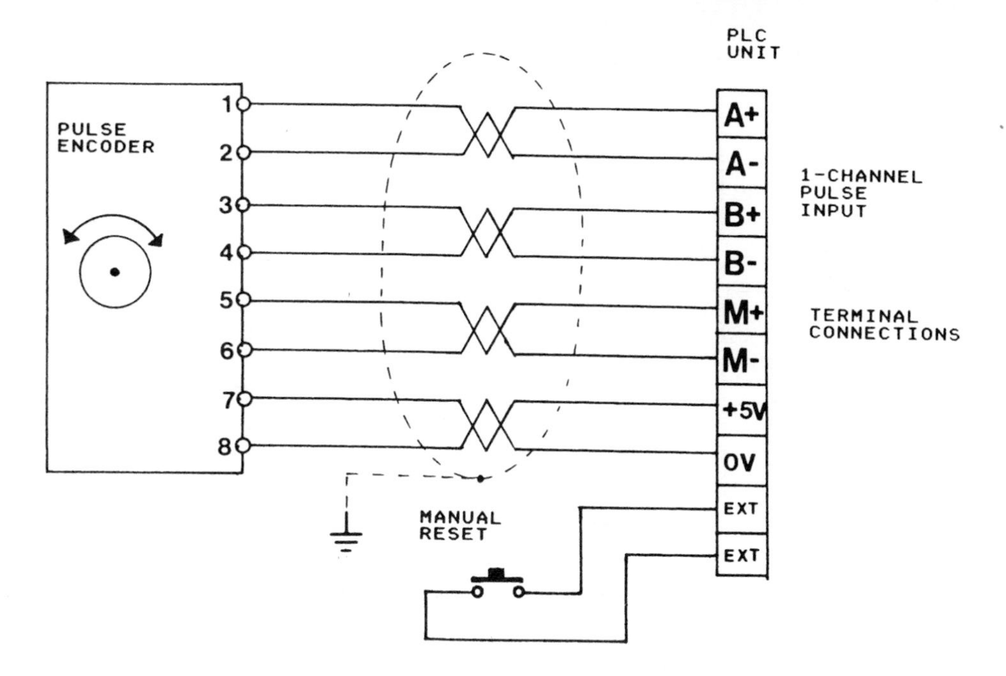

Figure 6.9 Position encoder connections

response to the signal transmitted from the PLC unit, the actuator mounted on the valve raises or lowers the valve stem. The valve plug is attached to the stem so that movement of the stem varies the opening of the valve.

The mass flow rate of the feedwater through the valve is determined by the opening of the valve and the valve opening is determined by the position of the stem. Therefore, a low current or voltage from the PLC to the actuator will result in a very low flow of feedwater. If maximum current or voltage is applied to the actuator then the valve will be fully open, resulting in a very large flow.

6.6 Electrical motors

Electrical motors, both a.c. and d.c., have evolved over the years to play a vital role in industry to convert electrical power into mechanical power. They are inherently straightforward to use, can be controlled by the inclusion of an electronic speed control unit and are relatively cheap to maintain and replace.

In Fig. 6.11 you will see a general layout of connecting a single-phase a.c. motor to the output of a PLC unit. The first relay connected to the PLC is called an interposing relay. Its function is to switch the low d.c. voltage to the main contactor coil which will in turn switch on the a.c. mains supply to the motor. This type of circuit layout will provide greater isolation between the PLC unit and the a.c. motor. You can also use this arrangement for d.c. motors.

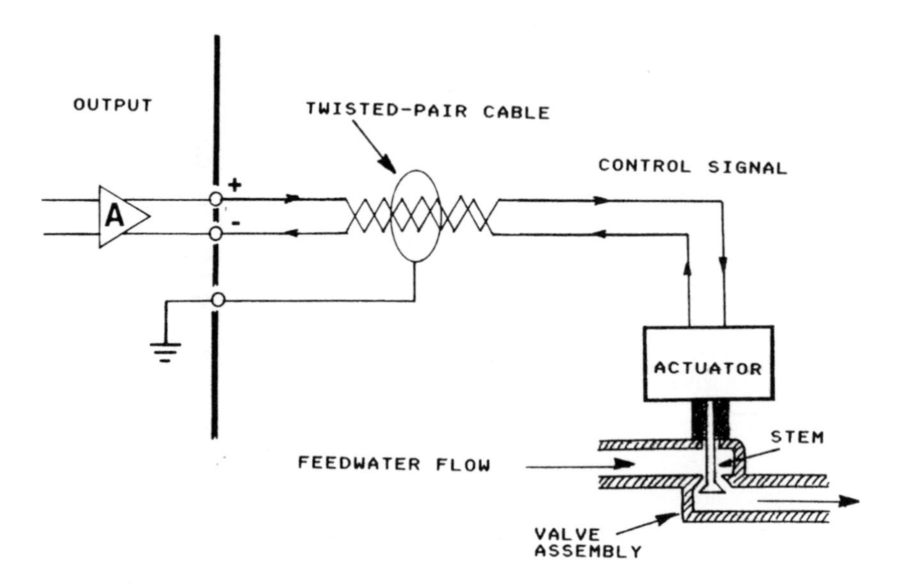

Figure 6.10 Principle of operation of servo valve

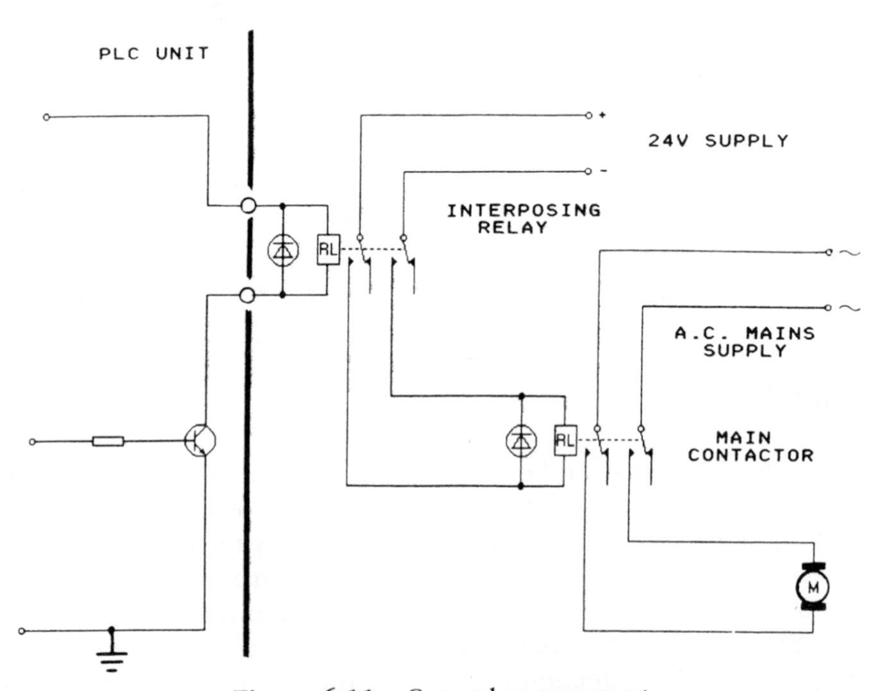

Figure 6.11 General arrangement

For a three-phase a.c. motor, another arrangement is shown in Fig. 6.12. A servo control unit is included and this provides a closed loop control as previously mentioned in Chapter 1. Where three-phase electrical supplies are available, this method of interfacing an electrical motor is generally preferred.

However, the installation of all electrical motors should be done in accordance with the manufacturer's rigid specifications. The importance of regular maintenance will ensure that the motors are kept clean, dry and in good working order. The ventilation grills and the air filters should also be kept clear of becoming obstructed in normal use.

6.7 Linear motors

A linear motor is basically an ordinary rotating motor which has been cut along one side and rolled out flat. There have been many applications for the linear motor such as opening doors and driving conveyor belts, etc., but they are now widely used for driving overhead cranes.

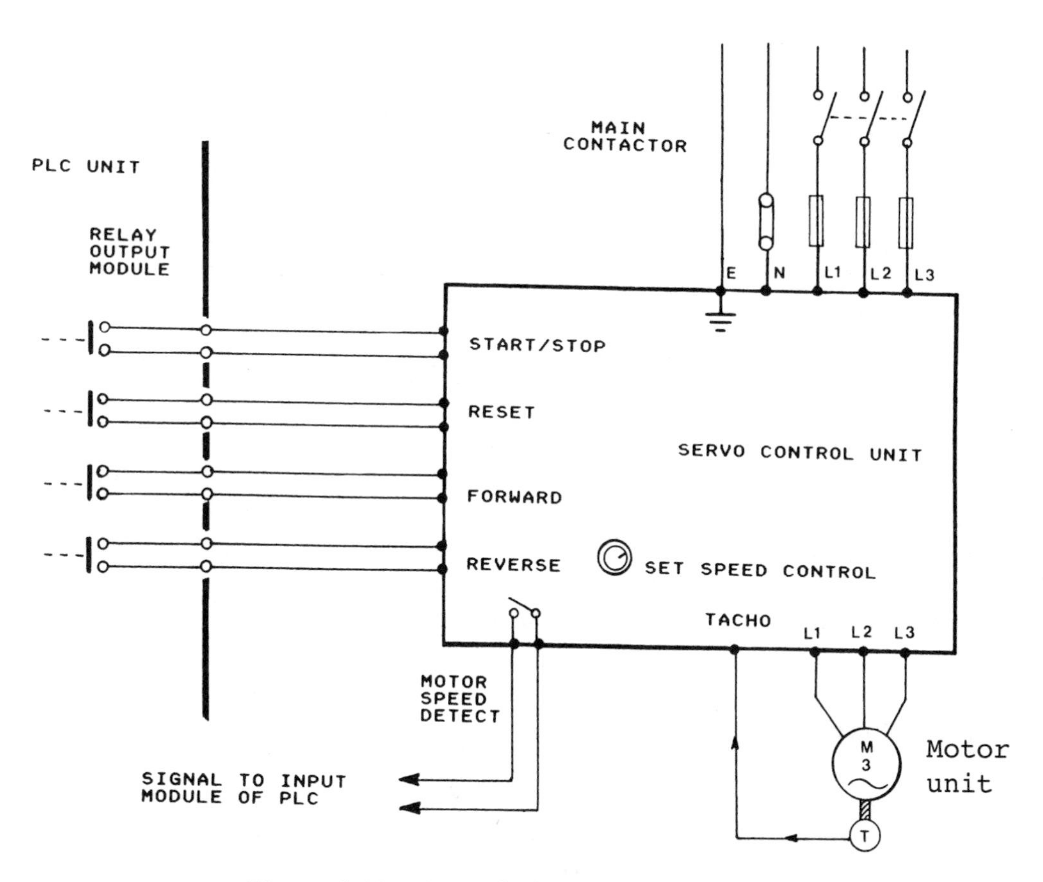

Figure 6.12 Control of a three-phase motor

In the application of the overhead crane, the beam which supports the crane becomes the secondary winding and a three-phase primary winding is attached to the frame of the motor which travels along the beam on wheels. This arrangement simplifies all the complications associated with rotary motors and their gears and drive shafts, resulting in a simplification of maintenance and repair as well as an improvement in reliability.

6.8 Stepping motors

The stepping motor is probably the most widely used in the control of machines for textiles, printing, heating, ventilation, process and automated industries. Other applications include paint spraying systems, industrial sewing machines, photocopying machines, labelling machines, printers, plotters and many more applications where accurate positioning and speeds are required.

Thus the stepping motor has the characteristic property of revolving in an exact number of individual steps and often a direct link between digital information and the incremental mechanical displacement.

The definition of the step is the process involving a rotation of the motor shaft by the step angle of the input control pulse. In each step position the rotor is held due to the electrical d.c. exitation of the windings which provides the holding torque for the motor shaft; see Fig. 6.13.

Stepping motors are available in many different designs. They are efficient, and high torques can be achieved with relatively small-sized motors. They can also be operated without the need for feedback, i.e. open loop control, as the steps are precise and accurate. However, for most high-performance applications, a servo system will be used to ensure accurate and reliable operation of machines and plant.

The stepping motor usually comes complete with a program and control unit, as well as a drive control unit. Refer to Fig. 6.14. A program is entered into the control unit for the stepping motor and all that is required of the PLC unit is to switch it on at the correct sequence and start the program. The inputs to the PLC unit are feedback signals to indicate either a limit position or that the correct speed has been achieved, and when the program has ended.

You will also notice that there is a manual override facility so that the stepping motor can be operated for testing and maintenance purposes.

6.9 Summary and test questions

In this chapter I have concentrated on analogue devices and you will have learned that the inputs and the outputs use converters. Hence, the inputs to the PLC will use analogue-to-digital converters and the outputs from the PLC will

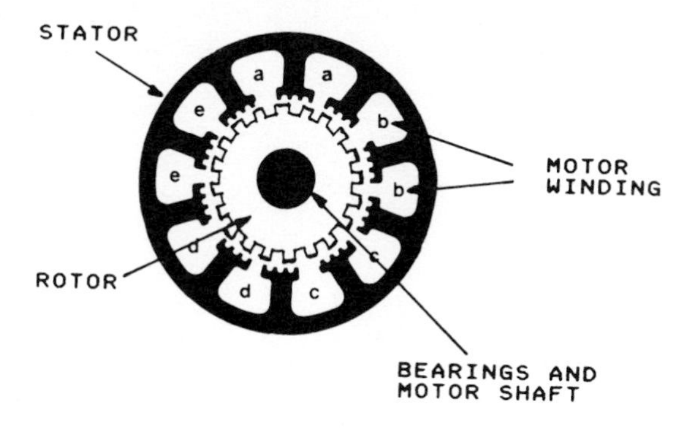

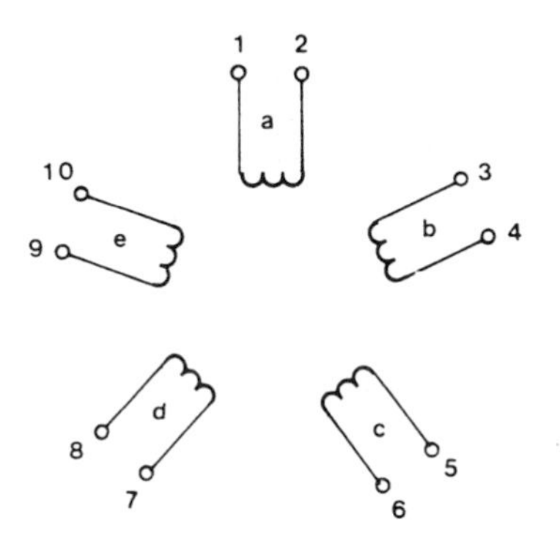

Figure 6.13 Cross-sectional view of a five-phase stepping motor

use digital-to-analogue converters. Thus interfacing with analogue input and output field devices is made easy.

I have tried to show some of the most popular interfacing techniques for PLCs, but there are many more which are beyond the scope of this book. Try and answer the following test questions.

1. Servo control can best be described as:
   (a) quasi-closed loop control
   (b) open loop control
   (c) closed loop control
   (d) sequential control.
2. What is an encoder device?
3. Why do we use twisted-pair cable for connecting up field devices?
4. A strain gauge is generally used to measure:
   (a) temperature
   (b) force
   (c) flow
   (d) position.

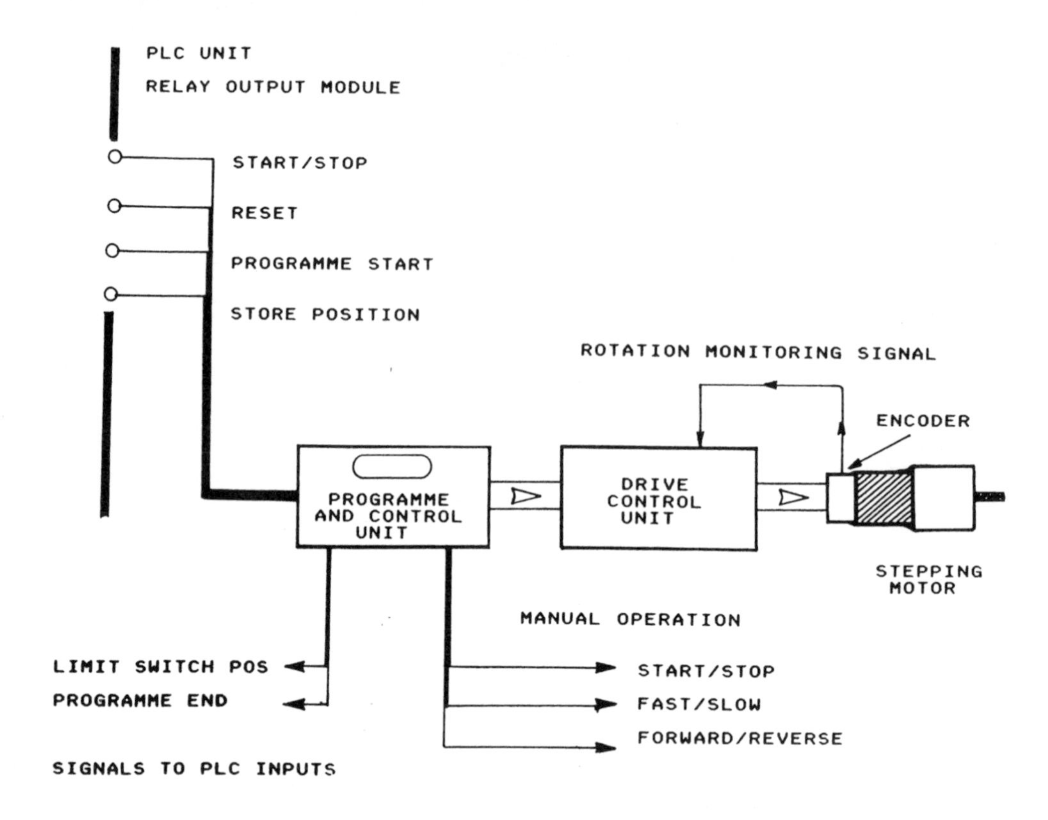

Figure 6.14 Stepping motor with servo system

5. What are the advantages of using a 4–20 mA current loop compared to a 0–10 V for the input signal to a PLC unit?
6. Why are stepping motors used instead of ordinary a.c. or d.c. motors for the control of machines?
7. Position encoders are used to:
   (a) measure temperature
   (b) provide feedback for rotation and limits
   (c) speed up the motors
   (d) control the flow of liquids.
8. What will be the resolution of a 12-bit analogue-to-digital converter.

7
Maintenance, Test and Inspection

7.1 Installation

The installation of any PLC unit must be done in accordance with the manufacturer's instructions. They will provide you with the details of the mounting arrangements, voltage requirements and a number of important specifications such as working temperature, dimensions, weight and environmental considerations.

First, you must select the proper type of enclosure for the PLC unit which conforms to electrical standards. This enclosure will protect the PLC unit as well as other devices from damage caused by contaminants in the air such as moisture, oil, dust and corrosive substances. It should also be large enough to allow easy access to the PLC unit and other associated devices and wiring for maintenance and testing. It is also important to locate the PLC unit in the enclosure away from any heat-generating equipment and provide adequate ventilation.

The enclosure will also give you protection from accidental contact of high voltages in normal use and prevent any mischievous or ill-considered operation or adjustment. It will provide security, safety and preservation of the equipment.

The earthing of the PLC and its associated equipment as well as the enclosure is very important. PLCs are solid state control systems, therefore good earthing will help to limit the effects of noise due to electromagnetic interference (EMI). The earthing path for the PLC and its associated equipment must be provided by an earthing conductor and a ground bus; refer to Fig. 7.1.

Earth connections should run from each of the devices to the ground bus direct. In other words each device must be grounded separately as shown. In addition to the earthing of all the devices inside the enclosure, a good earth must be provided to the enclosure.

7.2 Inspection and testing

The installation must be thoroughly checked prior to switching on the power supply. An improper connection can result in the equipment becoming dam-

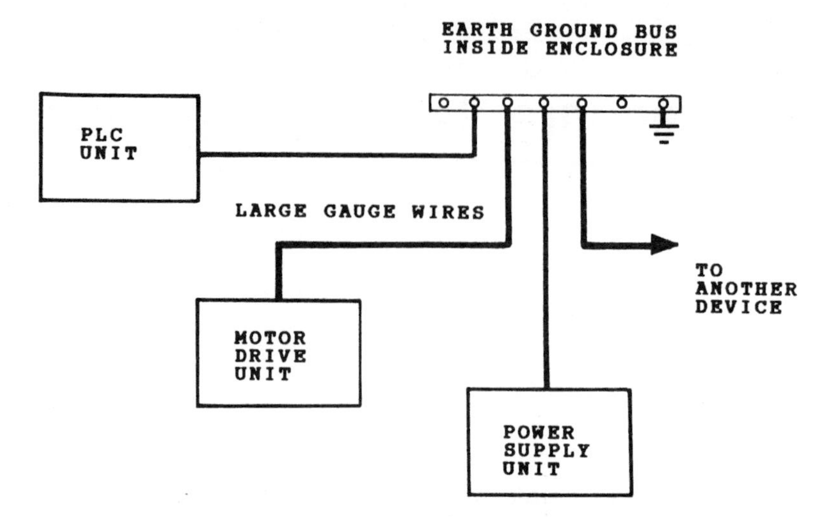

Figure 7.1 Earth grounding technique

aged. It is a good idea to use a report form as shown in Appendix 1 for recording purposes.

Make sure that the PLC and other devices in the system are securely mounted and free from obstruction. Check all the wiring and the connections, including the relays, emergency stop circuits and contactors. Inspect all power supply lines and measure the line voltage to be certain that it is correct for the requirements of the PLC and the associated equipment.

It will also be necessary to confirm that the installation complies with the manufacturer's drawings and specifications. This may seem a simple matter, but what appears to be simple on a drawing may turn out to be very difficult to appreciate on site.

When all the safety checks have been made and everything appears to be satisfactory, you can then power up the controller, load the PLC software with an appropriate program and then switch the [RUN/HALT] key on the controller to RUN mode.

If the system fails to operate, it is important to determine the cause of the problem. In some cases one problem may cause other faults to appear and it is a process of elimination to find the fault.

Start by checking the power supply module voltages and fuses. A blown fuse could mean that there is a short circuit either in the power supply module or in the system and control wiring.

It is also possible for the software to become corrupted or there may be some errors in the design of the ladder program. You will need to check the program thoroughly. For example, if the output coil device Y012 has been used more than once, then the output Y012 will not operate correctly.

The input and output modules can fail and most manufacturers will include status LEDs to give you an indication as to the condition of the PLC unit. For example, indicators could be allocated for the following:

- [RUN/HALT] key
- central processing unit
- input/output
- low battery
- communication port.

There are now table lists available for the self-diagnosis of the PLC unit. If the system does not operate correctly a displayed message will be generated, conveying to the user the nature of the problem, thus making it a valuable tool for troubleshooting software and hardware.

7.3 Maintenance activity

Periodically you should check the following items:

- power supply for the correct voltage and integrity of the wires and cables;
- input/output: visually check the wires, cables and terminals as they must not be damaged; all modules must be securely attached and mountings of terminal blocks must not be loose;
- PLC environment should be clean, meeting the specifications of the manufacturer regarding temperature, humidity, vibration and dust, etc;
- software program should be free from errors; you can compare it with the master program.
- Battery back-up option: some manufacturers will recommend that you change it once a year.

WARNING!

Isolate the power to the equipment before any inspection is done to the terminal connections, modules and cable wiring. This is to ensure personal safety and to guard against the possibility of damage to the equipment.

Safety working procedures are normal practice in all factories. Before any work is carried out all associated equipment must be isolated, preferably locked in the OFF position and the keys held by an authorized person in charge of the work. Safety notices should be displayed to indicate that the equipment is out of service and that the relevant personnel are informed in the factory.

If the work must be carried out on live equipment, then adequate precautions against the risk of electric shock and damage are required; for example, wearing of protective clothing such as rubber gloves and boots, using insulated tools and test probes. Also, if guards are removed for maintenance make sure that the machine is stopped first or is otherwise safe.

7.4 Calibration and recording

When the PLC installation is complete and the testing and inspection is complete it is a good idea to record all of your findings for future reference. Therefore, you need to have a system of record-keeping that will enable you to review and evaluate your findings. Calibration, in general terms, is to provide measurements traceable to national standards and carried out in accordance with the details of the specifications. Where applicable the calibration will be carried out with the manufacturer's test procedures.

Keeping records of all equipment, i.e. an inventory, will enable you to administer a calibration program at regular intervals. Today, it is usually done by using a database program held on a computer.

When the equipment has been identified by the database for maintenance and calibration, the service engineers will receive a print-out copy giving details of the type, building, area, last calibration date and other useful information; see Appendix 2. On completion of the service a calibration label is attached to the equipment with the date for the next servicing schedule.

The frequency of any service and calibration will be decided by the company after periodic reviews. Once a year is normal, but may need to be more frequent as some systems may be operating for up to 24 hours a day.

7.5 Emergency stops

It must be understood that PLCs are not fail-safe devices because they employ solid state components which are inherently unsafe. What this means is that they have an unpredictable failure mode, i.e. the internal components of the PLC can either go open circuit or short circuit. Therefore, the PLC is not used in the primary emergency stop circuit, but can be used to detect the source of the fault and indicate the information to either a plant mimic diagram or to a computer visual display unit.

Emergency stops must be hard-wired and must not depend on the software or the solid state logic of the PLC unit. See Fig. 7.2. The power to the main contactor should be removed immediately by the operation of one of the emergency stop push buttons. A secondary circuit provides the PLC with signals to indicate a fault condition. The circuit is said to be fail-safe; if any of the components (Fuse, ES1, ES2 or ES3) go open circuit the contactor will remain off.

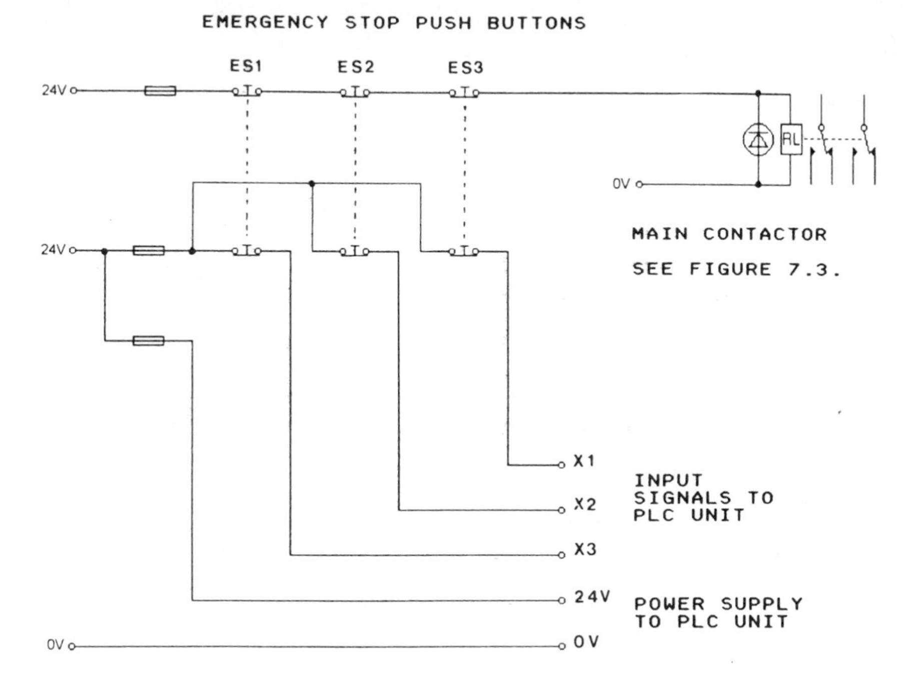

Figure 7.2 Emergency stop circuit

To provide further safety for an emergency stop situation, especially with three-phase a.c. motors, duplicate circuits are used. See Fig. 7.3. In this circuit two contactors are shown in series and their coils in parallel. This arrangement is very popular because relay contacts are not considered fail-safe. Imagine what would happen if a set of contacts became welded together. It is clear to see that the motor will continue to run even though one of the emergency stop buttons was operated. Therefore, dual circuits will provide some degree of extra safety. A ladder program is shown in Fig. 7.4 for an emergency stop indication and alarm. You may need to modify it slightly for your particular PLC model.

Futhermore, these duplicate safety circuits are now built by a number of leading manufacturers as a standard safety relay unit. Also, these safety relay devices will conform to British and European Standards for the safety of industrial machines and electrical equipment.

7.6 Health and safety

It must be understood that all electrical devices can constitute a safety hazard. It is therefore the responsibility of the user to ensure the compliance of the installation with any acts or bylaws in force.

The safety rules are covered by the *Health and Safety Act 1974* and the *Electricity at Work Act 1989*, which states that electrical equipment must be

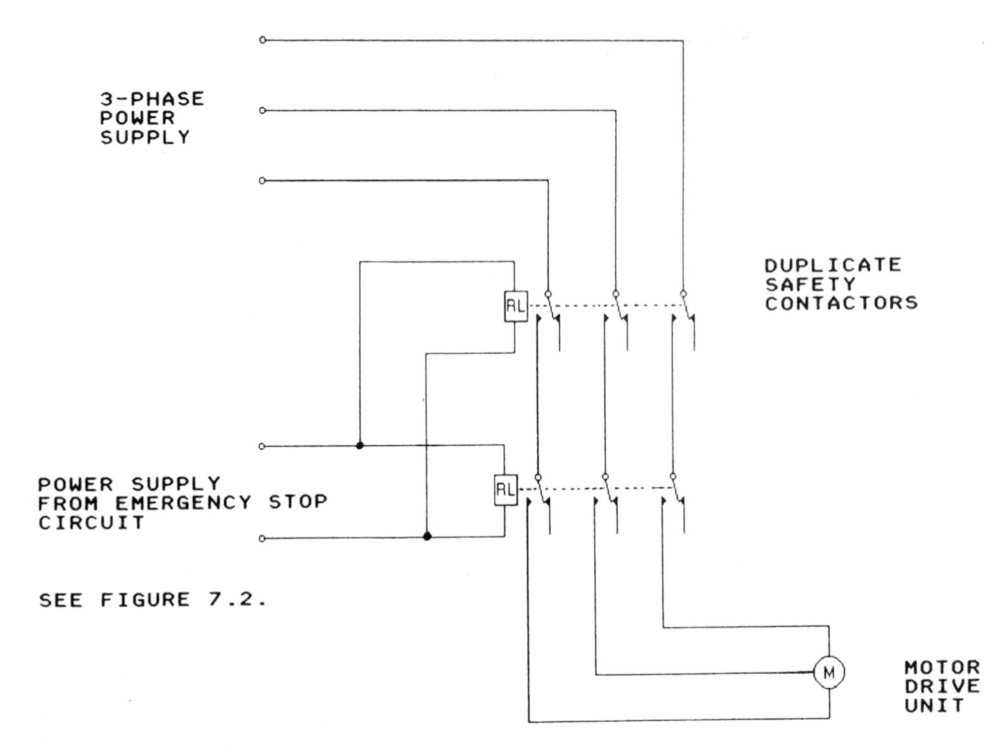

Figure 7.3 Emergency stop safety control circuit

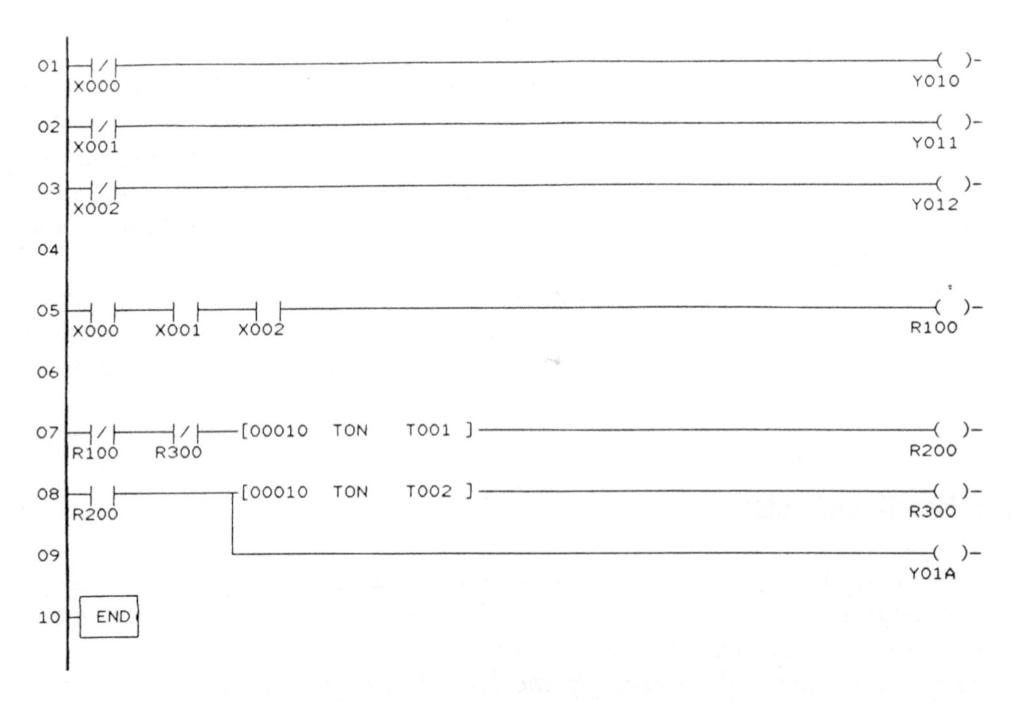

Figure 7.4 Emergency stop indication and alarm

designed, manufactured, installed, operated and maintained so as to avoid danger. The regulations relating to the installation and control of electrical and electronic equipment fall into two general categories:

- ❑ statutory or mandatory regulations which are backed by Act of Parliament;
- ❑ *Regulations for Electrical Installations*, issued by the Institution of Electrical Engineers (*IEE Wiring Regulations, 16th edition*).

The *IEE Wiring Regulations* are not backed by Act of Parliament; they remain as accepted recommendations. However, some of the regulations may relate directly to the statutory regulations and therefore may be enforced under that authority, e.g. the *Health and Safety at Work Act 1974*.

7.7 Summary and test questions

In this chapter I have discussed the implications of installation, inspection, testing, calibration and recording. Health and safety is also an important issue. It is a legal obligation in the UK for all companies to supply adequate information to ensure the safe use of machines and equipment under the *Health and Safety at Work Act 1974*. Therefore, safety devices must always be included to protect the machines from damage, as well as people from injury.

1. PLC modules can only be inserted or removed:
   (a) with the power supply switched on;
   (b) after the operation of an emergency stop;
   (c) with the power supply switched off;
   (d) when the PLC program has ended.
2. A PLC unit must be installed in an enclosure according to:
   (a) the local authority bylaws;
   (b) the manufacturer's specifications;
   (c) the health authority specifications;
   (d) the national standards specifications.
3. Explain why an emergency stop circuit must always be hard-wired.
4. Why is it necessary to have PLCs and other devices connected separately to a common earth or ground bus?
5. Calibration of PLC units and associated equipment should be done:
   (a) after two years
   (b) every week
   (c) every month
   (d) as required by company policy.

6. Who has the sole responsibility for the safety of electrical equipment and machines:
   (a) the government
   (b) the local authority
   (c) the manufacturer
   (d) the user.
7. If an emergency stop is operated, what should happen next?
   (a) the machine should continue to work
   (b) the operator must shut down the machine
   (c) the machine must stop immediately
   (d) an alarm will sound.

8
PLC Applications: Some Case Studies

8.1 A basic mixing station

Automatic mixing processes of liquids and other compounds in the chemical and food industries are very common. The basic mixing station can be seen as an automatic mixing process of two liquids for a specified time and then output of the final product to a storage tank; see Fig. 8.1.

The system consists of two level sensors which monitor the amounts of the two liquids flowing into the tank, three solenoid valves to control the flow of the two liquids and a motor connected to an agitator to mix the two liquids thoroughly.

This automatic mixing station will now require the following components:

Inputs to the PLC

Start push button	X000 – n/o
Stop push button	X001 – n/o
Level sensor No.1. (LS1)	X002 – n/o
Level sensor No.2. (LS2)	X003 – n/o

(n/o = normally open contact)

Outputs from the PLC

Valve No.1. (VA1)	Y010 – output coil
Valve No.2. (VA2)	Y011 – output coil
Motor starter (MS1)	Y012 – output coil
Valve No.3. (VA3)	Y013 – output coil

The start and stop push buttons are operated by the supervisor and these components are regarded as manual devices. The valves, sensors and motor are all called field devices and will enable the system to be operated by the PLC unit automatically.

The sequence of events for this basic mixing process will be as follows:

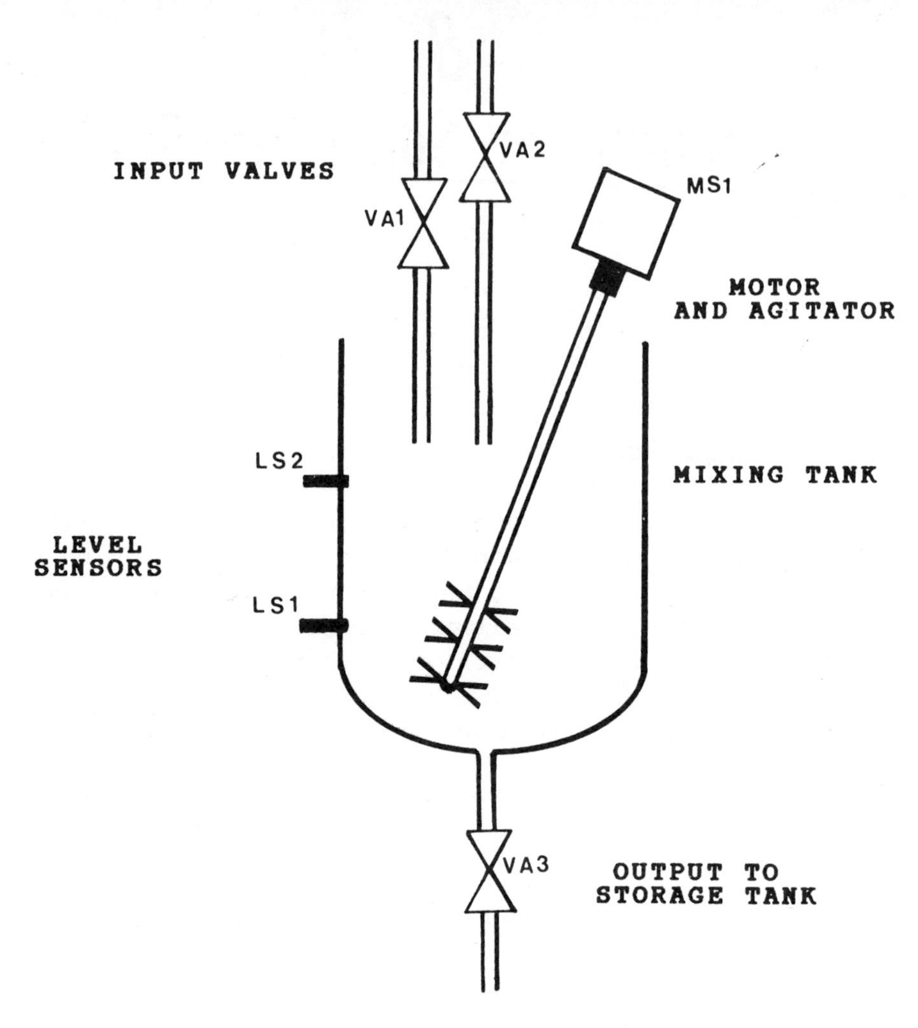

Figure 8.1 A basic mixing station

Step 1. Open valve 1 until level 1 is reached for the first liquid.
Step 2. Then close valve 1.
Step 3. Open valve 2 until level 2 is reached for the second liquid.
Step 4. Then close valve 2.
Step 5. Start the motor and agitate the two liquids for up to 20 minutes, or as specified.
Step 6. Then stop the motor.
Step 7. Open valve 3 for up to five minutes, or as specified, to empty the mixed product to a storage tank.
Step 8. Then close valve 3.
Step 9. Repeat or end the mixing process as required.

An example of a ladder logic program for this mixing station is shown in Fig. 8.2. Try it out, but remember you may need to modify it slightly for the PLC unit you are using, e.g. input and output codings, etc.

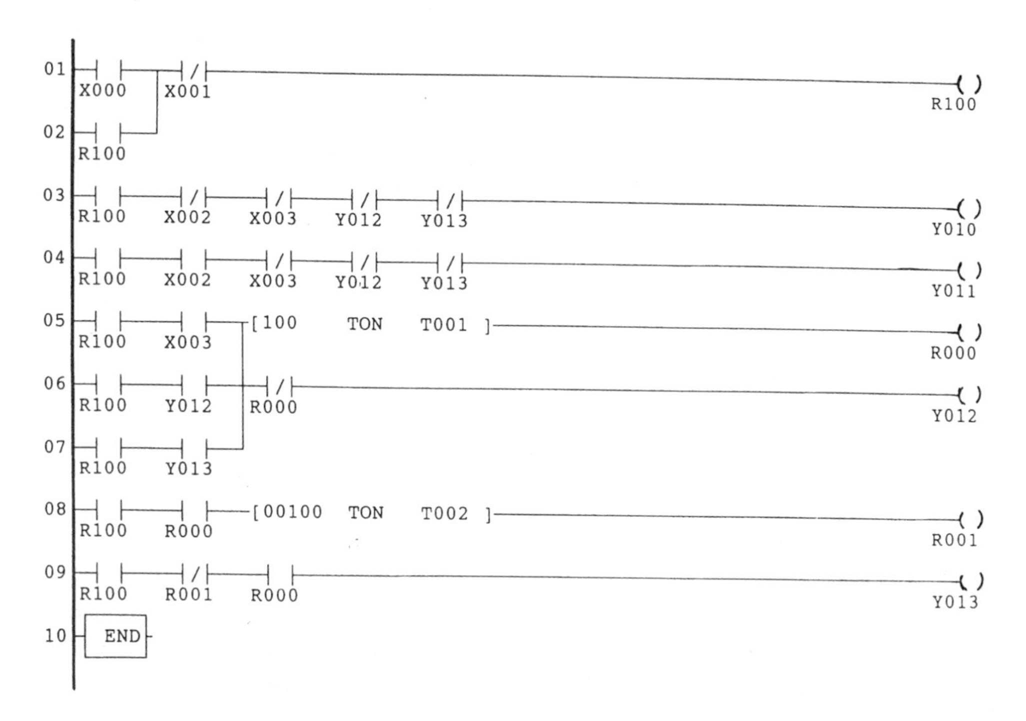

Figure 8.2 Ladder logic program for mixer station

8.2 Advanced emergency stop indication

In the previous chapter I showed you a ladder logic program for an emergency stop indication and alarm. However, if you wanted to know which device had tripped the process, it would be difficult to detect. Usually, what happens in a normal process, on a shut-down procedure, is that more than one fault will be indicated. This condition is due to other alarm and trip circuits being operated, for example a low flow to a process will result in a low pressure due to a valve being shut off in an emergency.

To provide a 'first-on' alarm, a different ladder logic program is required in order to detect which device had actually tripped the process first. A solution is shown in Fig. 8.3. You will notice quite a difference as this time the whole of the input card consisting of 16 inputs is used, i.e. there are 16 alarm or trip input circuits available.

On line 1, the input card is read as a word i.e. XW02, and the contents of this register are transferred to a register RW02. On line 2, the contents of register RW02 are compared with the EOR (EXCLUSIVE OR) register which has a con-

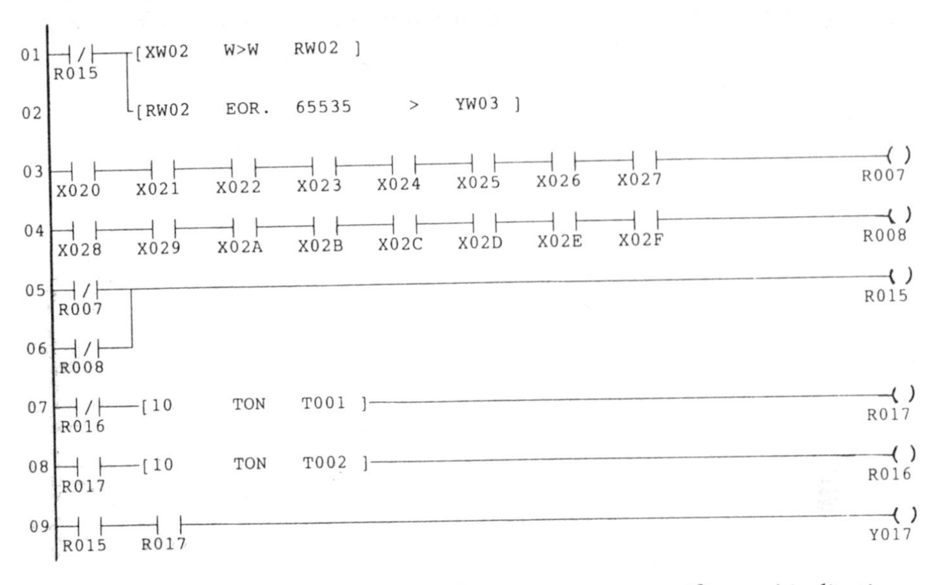

Figure 8.3 Ladder logic program for emergency stop 'first-on' indication

stant value of 65535. The output word is YW03 which will display the result when an emergency stop or trip has occurred. This arrangement will also latch, so that when the other devices operate the EOR function will ignore them.

The two timers T001 and T002 provide a switching pulse to the output Y017 to sound the alarm when R015 is activated by one of the inputs being operated in an emergency.

8.3 A three-stage conveyor system with batch control

Figure 8.4 shows the end of a production line, where the goods are tested and packaged. It is a very popular system and is used in almost every industry. Figure 8.5 is an example of the ladder logic program that is required for this system. The three conveyors (Y011, Y012 and Y013) start one at a time, with a 10 s interval, in order to reduce the starting surge on the electrical mains supply.

Y010 controls the flow of material on to conveyor 1. The material is to be tested by the sensor X005 and the outcome of the test will operate either the Y014 gate for a good artefact or the Y015 gate for a bad one. The Y016 gate will output a batch of artefacts on to the pallet, detected by the two sensors X001 and X002. When there are no artefacts going through the production line, the X006 sensor will go to its normally open state thus indicating to the PLC unit to shut down the line after a certain time limit.

This line can be automatically started when sensor X006 detects more artefacts coming from the production line. However, it is usual for the conveyors to be operating all of the time during a production run, as stopping and starting conveyors will consume more electrical energy.

The assignment list for this scenario is:

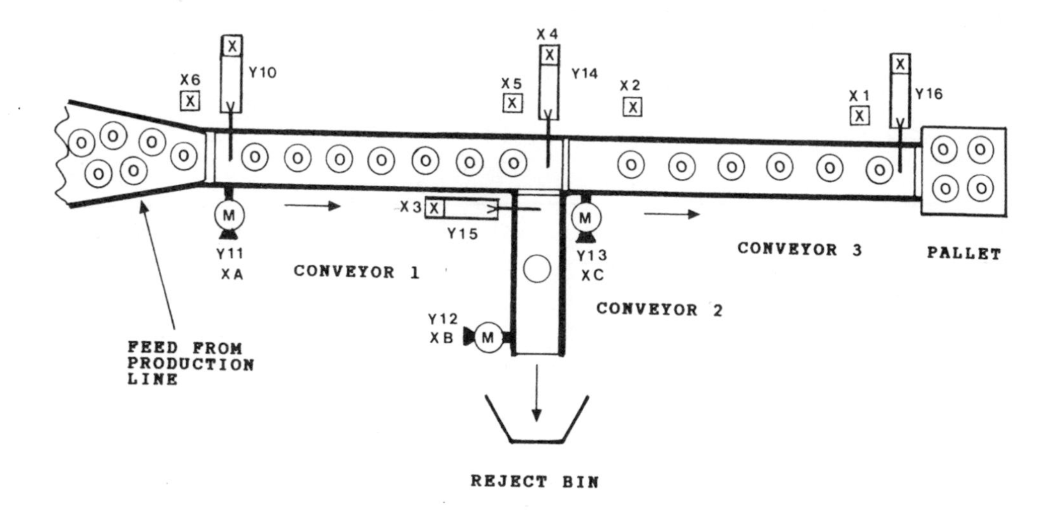

Figure 8.4 A three-stage conveyor system with batch control

Inputs to the PLC unit:

X1	Conveyor 3 empty sensor
X2	Conveyor 3 full sensor
X3	Reject gate open sensor
X4	Valid gate open sensor
X5	Goods quality sensor
X6	Goods input detect sensor
X7	Stop push button switch
XA	Conveyor 1 movement sensor
XB	Conveyor 2 movement sensor
XC	Conveyor 3 movement sensor
XD	Start push button switch

Outputs from the PLC unit:

Y10	Material feed gate to conveyor 1
Y11	Conveyor 1 motor
Y12	Conveyor 2 motor
Y13	Conveyor 3 motor
Y14	Valid gate to conveyor 3
Y15	Reject gate to conveyor 2
Y16	Output release gate from conveyor 3

Note: X00D = XD and Y014 = Y14. I have got rid of the leading 0's to simplify the diagram layout in Fig. 8.4.

```
X005   X00D   X007   R001                          R000
R000

X006   R000   Y011   R006   [100  TON  T002 ]      R001

R000                                               Y013
       X00C   [100  TON  T000 ]                    Y012
       X00B   [100  TON  T001 ]                    Y011

R000   X00A   R003   [30  TON  T003 ]              R002
              R002   [30  TON  T004 ]              R003
              R002   X002                          Y010

X002   [100  TON  T007 ]                           R004

R004   X001                                        Y016
Y016

X002   R004                                        R006
R006

X00C   X003   [20  TON  T005 ]                     Y014
Y014   X004

X00C   X004   [20  TON  T006 ]                     Y015
Y015   X003

END
```

Figure 8.5 Ladder logic program for a three-stage conveyor system with batch control

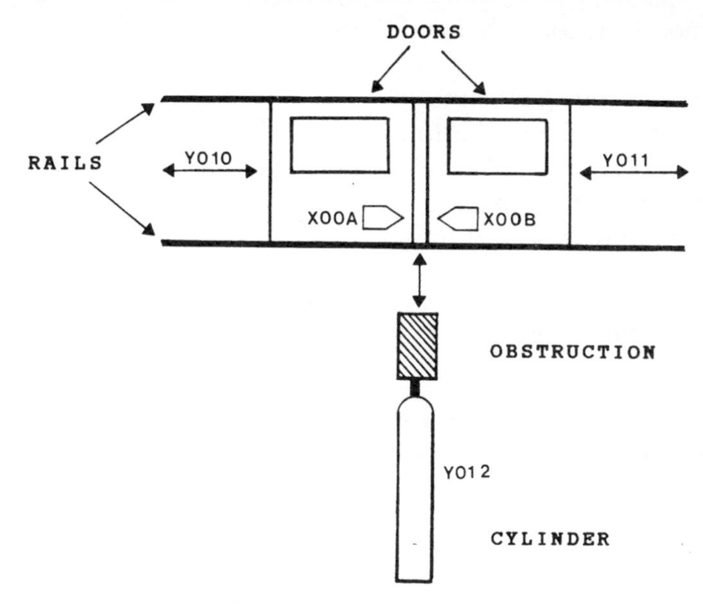

Figure 8.6 Automatic door control system

8.4 Testing an automatic door control system

A leading manufacturer has developed a sophisticated door system for trains, which improves passenger safety by preventing injury in the event of someone becoming trapped by the doors. The manufacturer needs to test the door control system automatically by opening and closing the doors about 1000 times. See Fig. 8.6.

Using a fully automatic test rig interfaced with a PLC unit, the doors must open and close 1000 times and on every fifth closing of the doors a pneumatic cylinder outstrokes to simulate an obstruction.

The cylinder will remain in the obstruction position while the doors attempt to close three times. After the third attempt the cylinder instrokes to allow the doors to close fully. If the doors fail to operate correctly at any time the test sequence should stop immediately.

The assignment list for this program is:

C01 Counter (1000) door operations
C02 Counter (5) cylinder operations
C03 Counter (3) reset

R001 Start/stop latching relay
R002 Relay output for end of sequence
R003 Relay output to reset cylinder
R010 Relay output to Y010 coil
R011 Relay output to Y011 coil

T001 Timer delay for left-hand door

T002 Timer delay for right-hand door

X000 Program start push button
X001 Program stop push button
X002 Switch to open and close doors
X00A Left-hand door sensor
X00B Right-hand door sensor

Y010 Left-hand door
Y011 Right-hand door
Y012 Cylinder

A ladder logic program for this test sequence can be seen in Fig. 8.7. X00A and X00B are photo-electric sensors which will prevent the doors from closing on detecting the obstruction (simulating a passenger).

8.5 Summary

The case studies I have discussed in this chapter are typical applications of using PLCs in industry. There are many more, but it would take another book

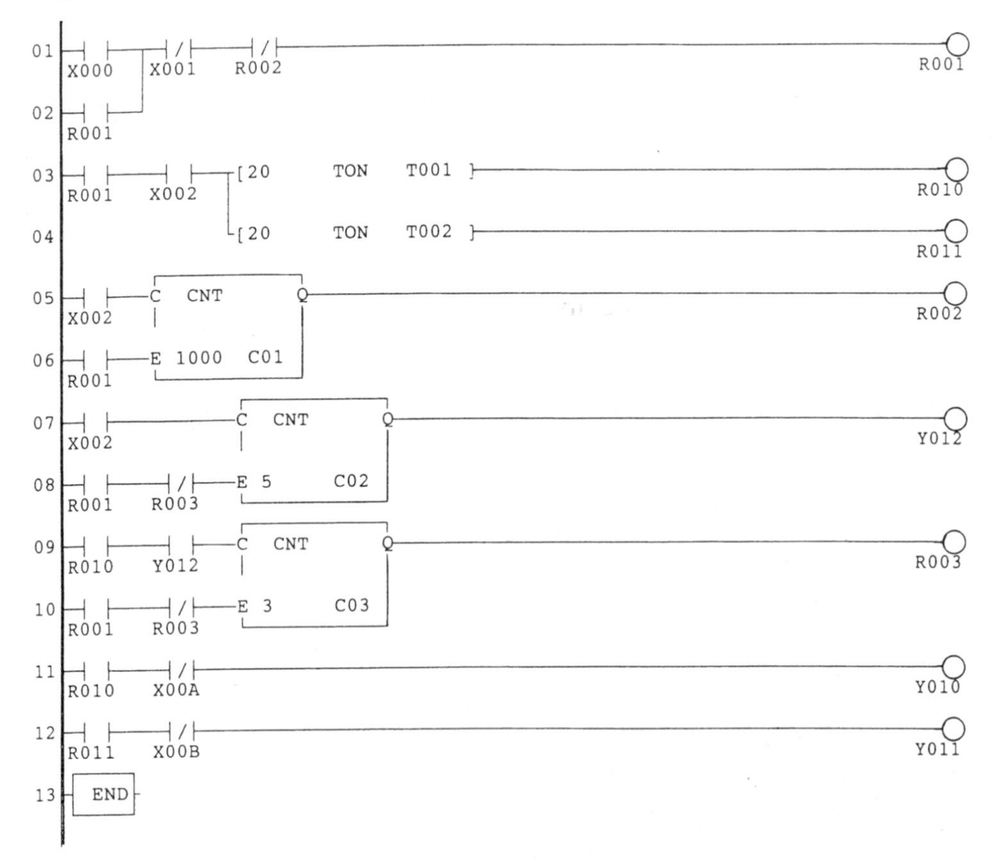

Figure 8.7 Test sequence program

to discuss them all. The solutions I have shown are typical; therefore, you may need to modify them for your own purposes.

When you decide to program your PLC unit, spend some time to plan the sequence on paper. This will also help you to keep a record for documentation purposes. Key in each subroutine separately and test it out before you start to include interrupts and interlocks. A complicated program at the start can become a nightmare to solve. Finally, make sure you save the program to a disk file.

9

The Next Generation: The Smart Distributed System for Factory Automation

9.1 Advanced technology

Technology is advancing in two major areas with respect to control systems for factory automation. The main drivers for these advancements are that users are demanding cost reductions and increased productivity. The two major advances being made are:

- intelligent bus-based networkable sensors and actuators are replacing conventional hard-wired devices;
- PCs are replacing the PLC in the same manner as PLCs replaced hard-wired relays.

9.1.1 THE SMART DISTRIBUTED SYSTEM

The **Smart Distributed System** is a bus-based control system for factory automation, enabling a complete portfolio of products to be connected on a bus. including digital and analogue I/O, sensors and switches, pneumatic valves, operator displays, motor starters, drives, PLCs and computers etc.

The system was designed by Honeywell. It is however totally open with many third party manufacturers developing and offering compatible products (see Fig. 9.1).

The extensive amount of cabling associated with conventional control systems and devices is eliminated using the Smart Distributed System as all the devices are simply connected to the system with a single twin twisted-pair cable. Two cores of the cable are used for communication and two for powering the internal circuitry of the devices. Each device has inbuilt intelligence and a unique address on the bus.

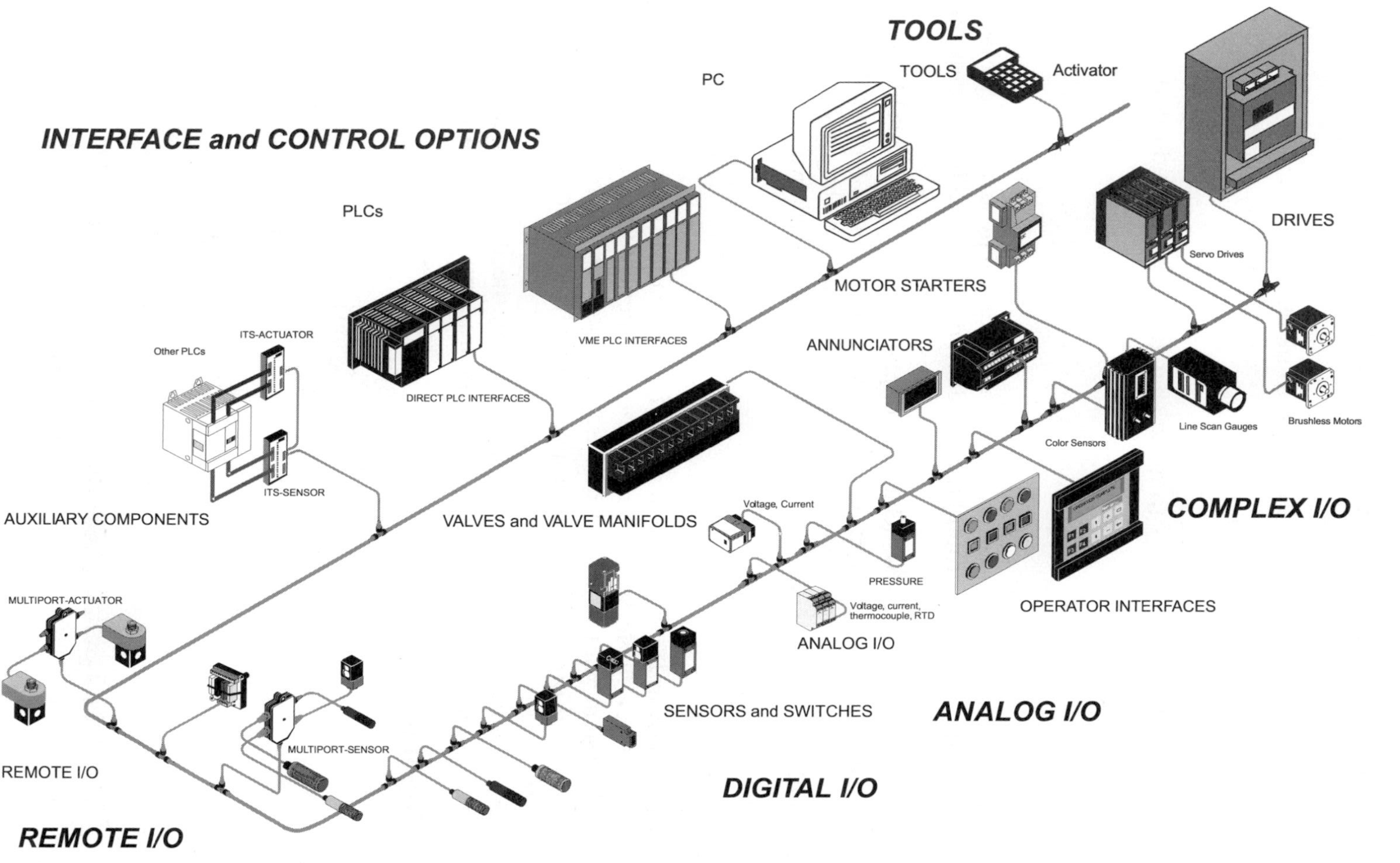

Figure 9.1 Overview of the Smart Distributed System

Dramatic installed cost reductions are made with:

- reduced design cycle times;
- reduction of hardware (cables, terminals, trunking, etc.);
- reduced build and test costs;
- reduced installation time;
- reduced on-site commisioning;
- increased flexibility (i.e. ability to make changes, move and add devices to the system).

9.1.2 TOPOLOGY

The topology of the bus is based on a trunkline, branchline arrangement (see Fig. 9.2). The trunkline runs around the installation with branchlines teeing off to devices.

Prederminated trunkline, branchline and pigtail cables are available in a range of lengths providing a large degree of plug'n'playability. Cable complying to the physical specification and preterminated pigtails can be used in conjunction with the user's own termination scheme as an alternative.

9.1.3 DIAGNOSTICS

As can be seen from Fig. 9.3, 75% of control system faults occur at the sensor and actuator level. Table 9.1 illustrates the real benefit of using the Smart Distributed System with its inbuilt extensive diagnostics. Current diagnostics capability is based on conventional programmable logic controller technology and standard sensors and actuators.

There is a dramatic increase, from 8% to 80%, in the ability to detect faults and failures using the system. The diagnostics are fundamental in increasing

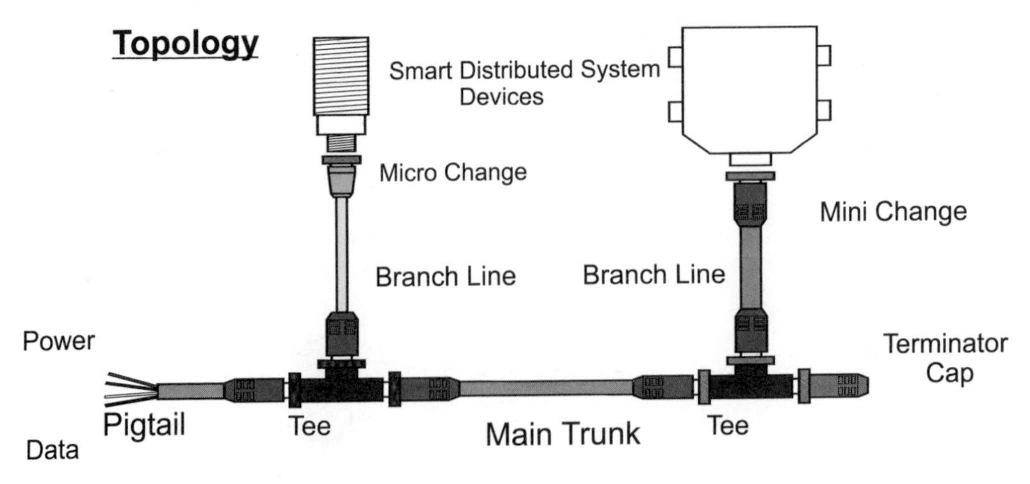

Figure 9.2 Topology of the system

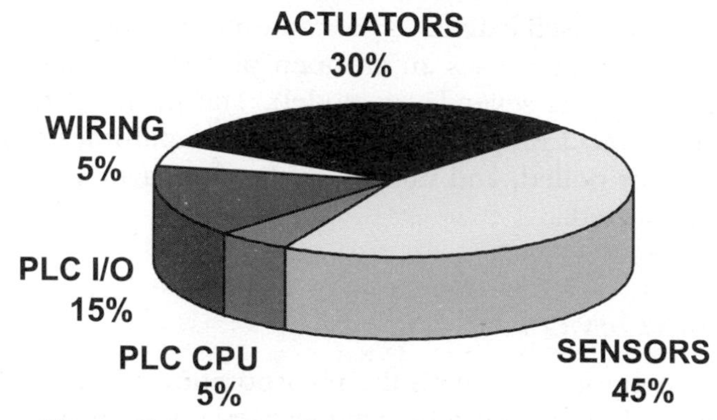

Figure 9.3 Control system fault distribution

productivity by highlighting the location of devices with abnormal operating conditions before failures occur, hence minimizing downtime and time to repair. Smart Distributed System installations have reported downtime reductions of up to 50%.

9.1.4 CONTROLLER AREA NETWORK (CAN)

Smart Distributed System devices use controller area network (CAN) processors as the data link layer (layer 2 of the ISO OSI seven-layer model). CAN was

Table 9.1 Fault detection

Fault source	Control system fault distribution	Current diagnostics capability	Sensor actuator bus diagnostics
CPU – parity error, processor, etc.	5%	100%	100%
I/O – power supply, input/output module failure.	15%	20%	95%
Wiring – broken shorted wire to sensor or actuator.	5%	0%	80%
Actuators – short circuit, overload, open circuit coil.	30%	0%	75%
Sensors – mechanical, blocked photo-eye	45%	0%	75%
Total	100%	8%	80%

used due to its widespread usage, availability from most chip manufacturers, robustness, non-destructive arbitration, advanced error management and small package size. CAN is used extensively for automotive onboard applications. The Smart Distributed System uses an an open protocol for the application layer (layer 7 of the ISO OSI seven-layer model). The protocol features user-selectable baud rates upto 1 Mbps, and peer-to-peer communications. Messages can be event driven or polled, and various application layer services and telegram lengths are supported.

9.1.5 INTELLIGENT DEVICES

The devices are empowered with the incorporation of CAN processors within them. Sensors and switches can be parametrized for the features shown in Table 9.2.

The device information shown in Table 9.3 is readily available over the bus. Smart Distributed System sensors and switches have extensive inbuilt diagnostics which monitor:

Table 9.2

Feature	Benefit
Normally open/closed	Reduction of inventory
On/off delay	Reduces load on control platform thus increasing system performance
Batch counting	Higher speeds of operation attainable; increasing productivity
Motion detection	Highlighting potential faults (jams, etc.)
End of life (No. of operations)	Preemptive maintenance

Table 9.3

Feature	Benefit
Prefailure diagnostics	Highlighting potential problems or failures
No. of power cycles and operations	Monitor productivity and speed
Bus diagnostics	Highlighting device and system faults
Elapsed time	Preemptive maintenance
Vendor information: manufacturer, device type, part number, date of manufacture, serial number	Easily accessed over the bus; simplifies documenting and spares ordering

1. Change in operating conditions

Proximity sensor	Target too close or too far
Photo-electric sensor	Marginal gain caused by misalignment or dirty lens
Limit switch	Slow lever return, indicating mechanical fault
All devices	Bad or missing device

2. Environmental conditions
   Temperature
   Moisture ingress
   Nominal current

The devices report on any abnormal operating conditions or failures immediately pin-pointing the location of potential problems.

9.2 PC control

PLCs in factory automation can no longer compete with the PC ability to improve efficiency and productivity, and its flexibility to adapt to changing needs. The transistion from PLCs to PC-based control throws off restrictions that were taken for granted until now.

Open systems. Users resent being tied to vendor-specific proprietary control platforms (i.e. PLCs) and are now moving towards open architecture PC control. Industrialized PCs are a totally open control platform, allowing the user to interchange hardware components and select the preferred supplier from a wide range of manufacturers. PC control offers a wider range of programming and development tools.

Increased functionality. Dramatic increases in control system functionality and flexibility are introduced with PC control, allowing easier integration of other familiar Windows-based applications such as spreadsheets, databases and user-specific applications for management and reporting.

Networking of PC controllers to management information systems is far easier and less costly than PLC systems with the standardization of TCP/IP and ethernet. PC control offers enhanced diagnostics and error-handling capabilities.

9.2.1 INTEGRATED PC CONTROL

Honeywell's integrated PC-based control platform provides further enhancements to the intelligence of the Smart Distributed System (see Fig. 9.4).

The industrialized PC connects to the Smart Distributed System by means of an interface card that simply plugs into the standard ISA backplane, each card

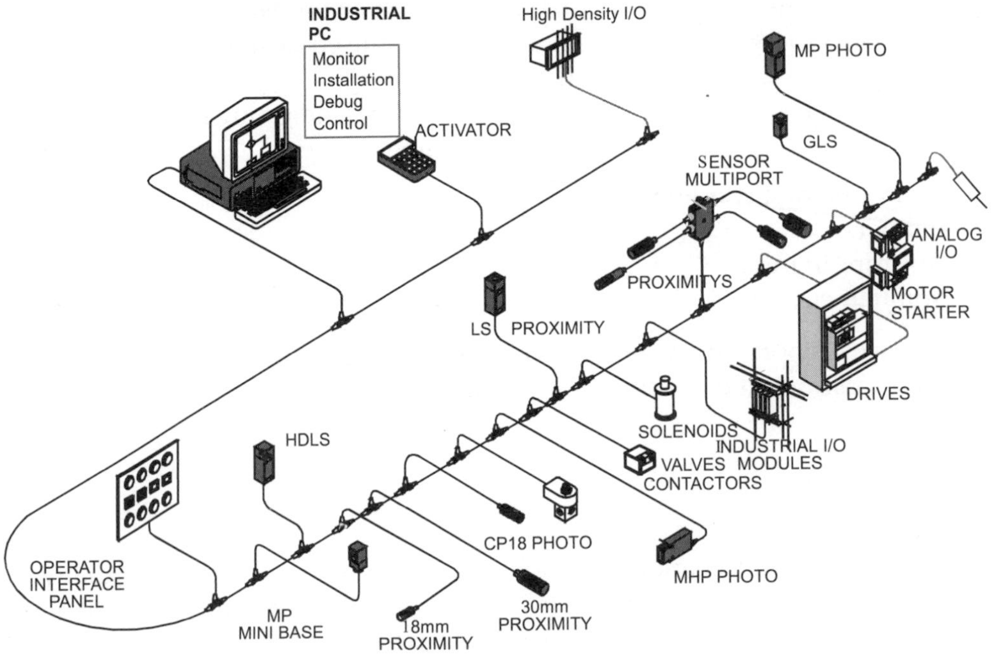

Figure 9.4 PC control

providing connection to two buses. Up to six such cards can be used with the PC control platform.

Integrated control enables network design and management, control program editing, real-time control and operator interaction to be performed on a single PC. This provides significant hardware cost savings over that of conventional PLC technology where additional programming workstations and human–machine interface (HMI) units are used in conjunction with the controller.

The integrated control software suite consists of the following applications:

- Smart Distributed System bus builder
- device manager
- control designer, maintainer, and runtime
- HMI builder and runtime.

Network design. The Smart Distributed System bus builder is a software tool to aid the design of the bus following the design guidelines. System documentation is also created with the bus builder software (Fig. 9.5).

Network management. The device manager is a software tool used to easily configure devices, allowing the configuration of all device parameters and attributes of the entire network from a single point. The software is used for monitoring to troubleshoot the network and devices.

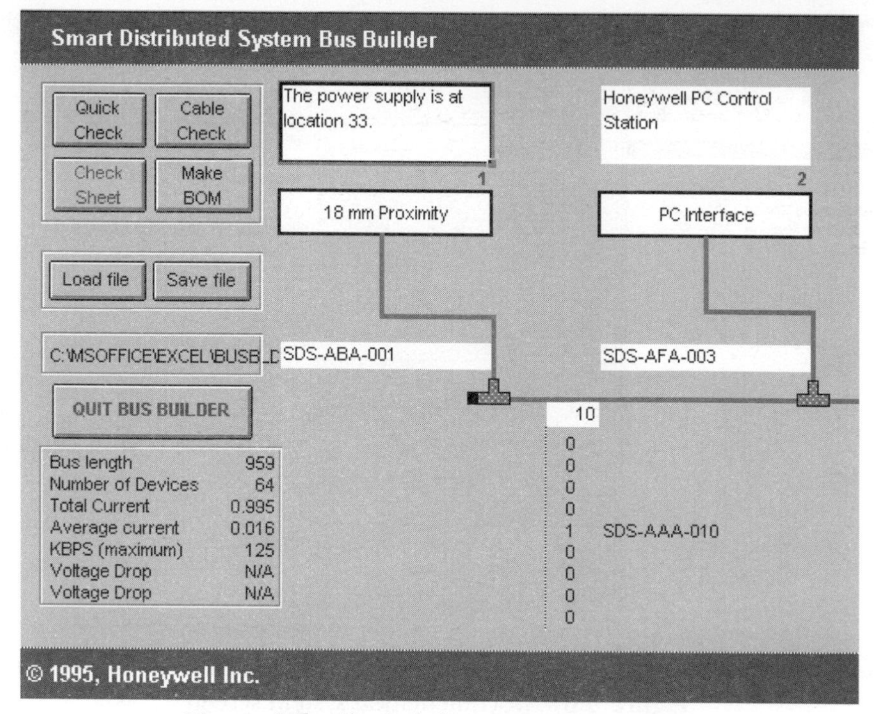

Figure 9.5 Typical device manager screen

The task of documenting the system is simplified as the software generates electronic system documentation, detailing all information of the devices on the network (Fig. 9.6).

Control program logic editor. Control program development is carried out in IEC1131-compliant flowchart and relay ladder logic with the Windows-based control program editor. Program development time is reduced due to the standard windows 'point and click' functionality allowing the programmer to copy, paste, etc., modules of the program. The subprogram facility allows a very structured approach to be taken to the overall control strategy, simplifying program design and reducing the time necessary to make changes.

Progamming is simplified by using flowchart programming as it produces a solution to the problem. Flowchart programming produces a structured, well-defined, self-documented system. Flowcharts have other benefits:

- they are totally independent of the hardware platform, as the desired sequence of events is simply described within standard flowchart symbols, hence no specialist knowledge of the controller internals is required;
- program logic can easily be read and understood by people other than programmers.

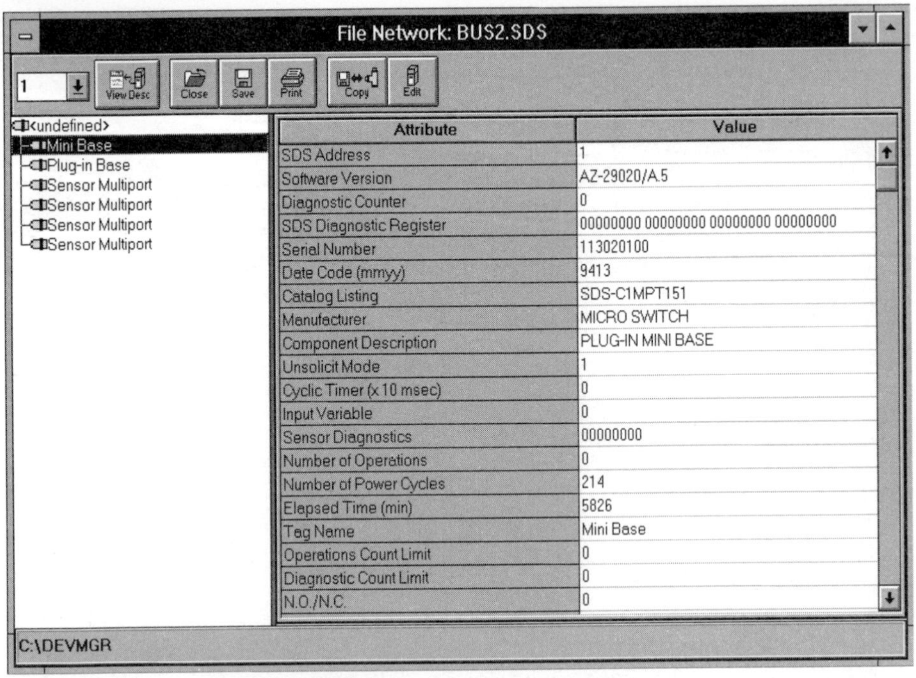

File Network: BUS2.SDS

1 | View Desc | Close | Save | Print | Copy | Edit

<undefined>
- Mini Base
- Plug-in Base
- Sensor Multiport
- Sensor Multiport
- Sensor Multiport
- Sensor Multiport

Attribute	Value
SDS Address	1
Software Version	AZ-29020/A.5
Diagnostic Counter	0
SDS Diagnostic Register	00000000 00000000 00000000 00000000
Serial Number	113020100
Date Code (mmyy)	9413
Catalog Listing	SDS-C1MPT151
Manufacturer	MICRO SWITCH
Component Description	PLUG-IN MINI BASE
Unsolicit Mode	1
Cyclic Timer (x 10 msec)	0
Input Variable	0
Sensor Diagnostics	00000000
Number of Operations	0
Number of Power Cycles	214
Elapsed Time (min)	5826
Tag Name	Mini Base
Operations Count Limit	0
Diagnostic Count Limit	0
N.O./N.C.	0

C:\DEVMGR

Figure 9.6 Documentation system screen

The relay ladder logic and flowchart languages can be combined with flowchart functions calling blocks of ladder for execution.

The control program is compiled after completion; error checking is carried out during compilation (Fig. 9.7). An error window describes any programming errors that occur within the control program, the exact location of which is indicated by clicking on the error description. The cursor then moves to the problem within the program. Program development time is reduced with the error-checking functionality.

The control program can be tested without connection to any field devices by using the simulated I/O driver and simulation program facility. Simulation programs are flowcharts that emulate the status of the field devices by allowing the status of inputs to be written.

The integrated control maintainer allows testing of the program. The execution of the flowchart is highlighted on the screen, allowing the user to monitor program execution. Furthermore, I/O, variables and registers, etc., can be monitored and their values forced with the control maintainer.

Integrity and reliability of the control program during execution is guaranteed as the compiled control program runs under a real-time operating system totally independent of DOS and Windows. This ensures that the control program will continue to execute even if DOS or Windows crashes.

The control program can easily exchange data with other Windows-based programs, such as databases, spreadsheets and user-specific applications used for reporting, by using the dynamic data exchange (DDE) facility.

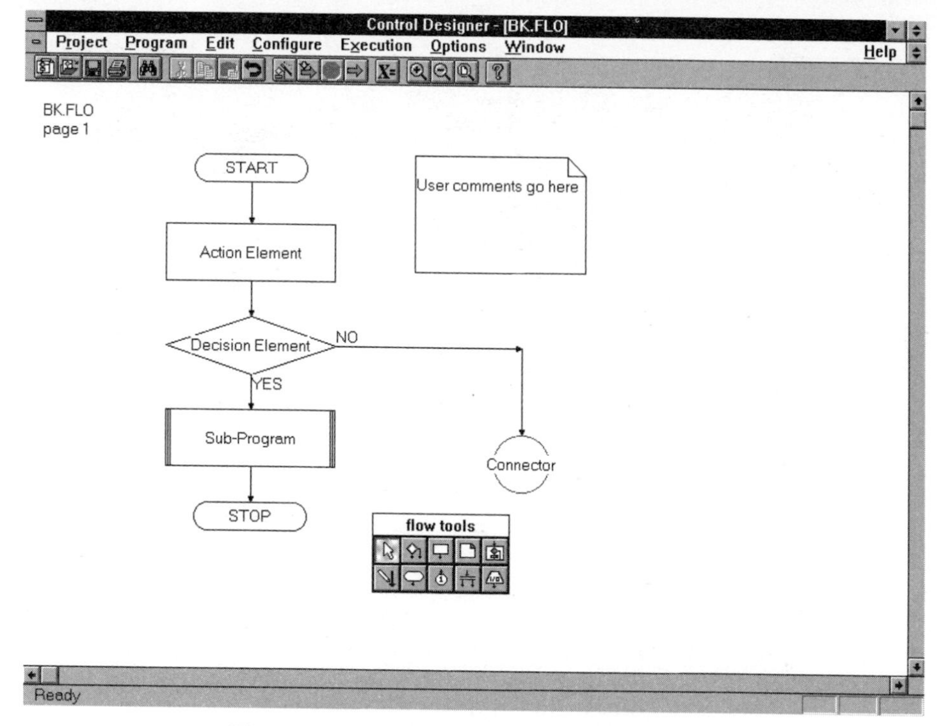

Figure 9.7 Typical control editor screen

Human–machine interface (HMI). The HMI enables complete visualization and interaction between the operator and the machine. The machine is graphically represented on the screen (Fig. 9.8). The operator is able to control the operation of the machine using graphical user controls (typically pushbuttons, sliders, data entry points etc). The graphical representation of the machine is continually updated as events occur. The HMI is used to display valuable information from smart devices.

The graphical user screens are easily created with the Windows-based HMI builder. The integration of the HMI screens and control program is seamless as they share a common tag database.

Integrated HMI diagnostics unlock the true power of intelligent devices, providing additional process information, predictive maintenance and they assist the user in self-diagnosis, all leading to major productivity increases.

Networking to other open higher level architectures (e.g. Ethernet) is carried out with the HMI, providing information to management systems. A very large control system would typically consist of many industrialized PCs, each one connected to a number of Smart Distributed buses, networked on Ethernet with a bridge connection to the higher level management system.

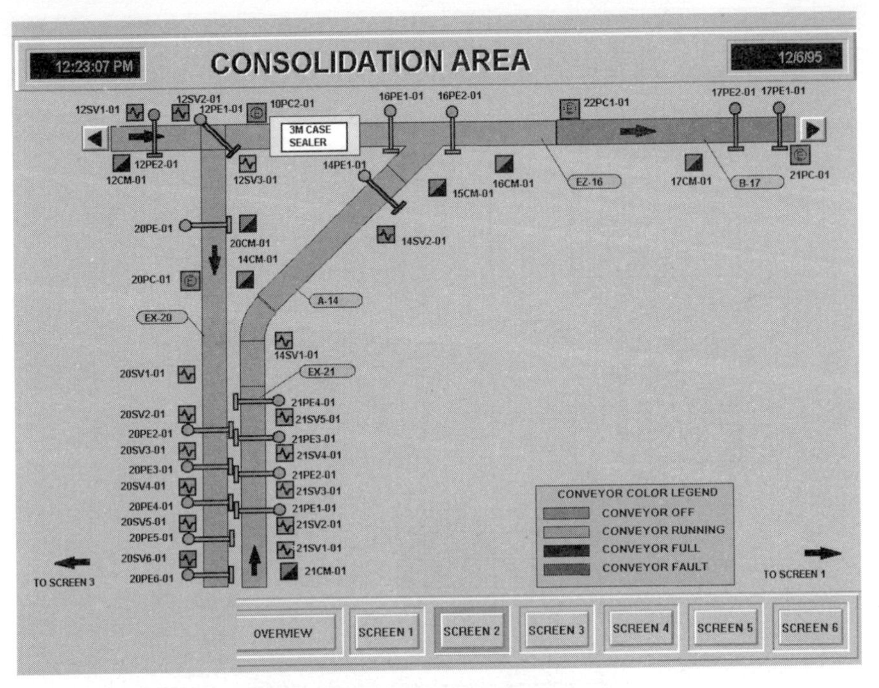

Figure 9.8 Typical HMI screen

9.3 Migration of the Smart Distributed System

The system is engineered to accommodate the evolution of control system requirements in the search for increased productivity, a standard for future control system architecture, and provides the framework for a host of advanced control technologies such as peer-to-peer communications, enabling devices to communicate directly with each other.

APPENDIX 1
Inspection and Maintenance Report Form

Checklist for industrial equipment

DATE:

ENGINEER:

CRITERIA:	REMARKS
LOCATION OF SYSTEM	
SYSTEM TYPE MODEL AND MANUFACTURE	
DRAWING NUMBERS AND RELATED DOCUMENTS	
JUNCTION BOXES	
CABLES CORRECTLY LABELLED AND CONNECTED	
TERMINAL CONDITION	
UNUSED CABLES/TERMINATED	
GOOD EARTH BAR	
CABLE GLANDS AND WEATHERING SEALS	
MOUNTINGS AND ENCLOSURES	
GENERAL COMMENTS	

APPENDIX 2
Calibration Record Sheet

AREA	BUILDING	TYPE	MAKE	SERIAL NO.

DATE	DESCRIPTION	SIGNATURE

FORM MC001 – Issue No.1

APPENDIX 3
Typical DOS Commands

FORMAT A:	To format a floppy disk
DISKCOPY A: B:	To copy a disk
CHKDSK A:	To check a floppy disk
CHKDSK A:/F	To fix errors on the disk
CHKDSK A:/V	To display error messages
DIR	To display files
DIR/P	To display files by page
DIR/W	To display all the files on the screen
MD or MKDIR (name)	To create a subdirectory
CD or CHDIR (name)	To change the directory
COPY A:*.*	To copy all the files from disk A: to hard disk C:
COPY A:*.* B:	Copy all files from disk A: to disk B: This is assuming that you have two disk drives. Otherwise, the computer will command you to change disks during the copy procedure
EXIT	To return to a previous command program

APPENDIX 4
Toshiba M20/40/EX100 Programmable Controllers

Figure A4.1 Typical PLC units

The Toshiba M20/40/EX100 programmable logic controller has a high speed program execution of 0.9 μs per contact instruction to meet control system requirements. The M20/40's program is written in a user-familiar ladder diagram with function block and can process 16-bit based high-performance instructions, such as data manipulations, arithmetic functions, logical operations, and special instructions. All I/O modules for the EX100 can be used with the M20/40 by using an expansion rack. The following I/O modules are available.

- 16 pts d.c. input
- 32 pts d.c. input
- 16 pts a.c. input
- 12 pts relay output
- 8 pts relay output
- 16 pts d.c. output
- 32 pts d.c. output
- 12 pts a.c. output
- 4 ch analogue in (8-bit)
- 4 ch analogue in (12-bit)
- 2 ch analogue out (8-bit)
- 2 ch analogue out (12-bit)
- 1 ch pulse input
- 1 axis position control
- ASCII/BASIC
- TOSHLINE-30 (wire)
- TOSHLINE-30 (optical)

The M20/40's built-in EEPROM and capacitor retain the program and data without need of a back-up battery. The clock-calendar function (year, month, day, week, hours, minutes, seconds), can be used for performing scheduled operations, data gathering with time stamp, and the like. The M20/40 has one port (RS-485) for data linkage to a computer. By using this computer link function, an operator interface or a supervisory system can be supported. On-line (during a run) program changes are possible. Effective program debugging and simulation are available without stopping operation.

The M20/40 are compact all-in-one box-type programmable controllers suitable for small-scale control applications. Programming instructions and peripherals for the M20/40 are fully compatible with the EX100. EX100 I/O modules can also be used with the M20/40.

The M20 has 12 inputs and 8 outputs. The M40 had 24 inputs and 16 outputs. The following expansion units and expansion racks are available for I/O expansion:

- 20 point expansion unit (20 exp)
  12 inputs and 8 outputs
- 40-point expansion unit
  24 inputs and 16 outputs

- ❑ 2-slot expansion rack
  2 I/O modules can be mounted
- ❑ 4-slot expansion rack
  4 I/O modules can be mounted

One additional expansion unit or expansion rack can be connected to the M20/40. The following combinations are available.

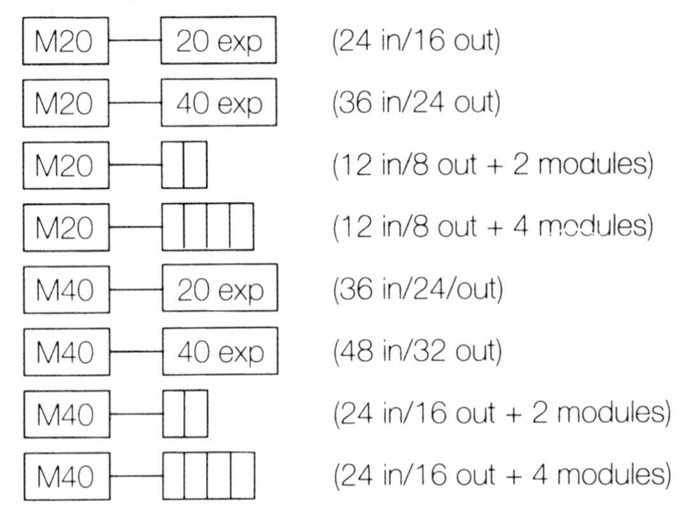

Figure A4.2 Input/output combinations

M20/M40 GENERAL SPECIFICATIONS

Power supply voltage
100 to 240 V a.c. (+10%, –15%), 50/60 Hz
24 V d.c. (+20%, –15%)

Power consumption
50 VA or less for a.c. power supply
22 W or less for d.c. power supply

Retentive power fault
10 ms or less

Withstand voltage
1500 V a.c., 1 minute

Ambient temperature
0 to 55°C for operation
–20 to 75°C for storage

Ambient humidity
20 to 90% RH, non-condensation

Noise immunity
1000 Vp-p/1 μs, NEMA ICS3–304

External dimensions (mm)
M20/20 exp: 240(W) × 125(H) × 78(D)
M40/40 exp: 320(W) × 125(H) × 78(D)
2-slot rack: 98(W) × 132(H) × 107(D)
4-slot rack: 168(W) × 132(H) × 107(D)

Approximate weight
M20/20 exp: 1.5 kg
M40/40 exp: 2.0 kg
2-slot rack: 0.5 kg
4-slot rack: 0.9 kg

CPU MODULE

Standard

- 4k program memory
- Write password protection
- On-line programming
- Built-in EEPROM and capacitor (battery is optional)
- High scan speed
  0.9 μs/contact
  110 μs/16 bit addition

Enhanced same as standard, plus:

- Built-in RS485 computer link
- Real time clock/calendar

POWER SUPPLY MODULES

- Types
  - 100/120 V a.c.
  - 200/240 V a.c.
  - 24 V d.c.
- Features
  - Provides power to CPU
  - External 24 V d.c.–0.5 A power
  - Run status contact

HIGH PERFORMANCE SOFTWARE

- ❑ Easy to use ladder diagram programming
- ❑ Function blocks can be inserted into ladder diagrams
  - ❑ Maths (double-length registers)
  - ❑ Data move (register-table, etc.)
  - ❑ Logic (AND, OR, etc.)
  - ❑ Comparison (<, =, >)
  - ❑ Trig. (sin, cos, etc.)
  - ❑ Limit
  - ❑ Special (FF, SR, sequencer, function generator, min., max., avg., etc.)
  - ❑ Immediate I/O update
  - ❑ EEPROM access (V2.1 or after)

STATUS AND CONTROL

- ❑ Run
- ❑ CPU
- ❑ I/O
- ❑ Com link
- ❑ Key switch access
- ❑ Station selector

I/O MODULES

- ❑ 16 pts d.c. input
- ❑ 32 pts d.c. input
- ❑ 16 pts a.c. input
- ❑ 12 pts relay output
- ❑ 8 pts retay output
- ❑ 16 pts d.c. output
- ❑ 32 pts d.c. output
- ❑ 12 pts a.c. output
- ❑ 4 ch analogue in (8-bit)
- ❑ 4 ch analogue in (12-bit)
- ❑ 2 ch analogue out (8-bit)
- ❑ 2 ch analogue out (12-bit)
- ❑ 1 ch pulse input
- ❑ 1 axis position control
- ❑ ASCII/BASIC
- ❑ TOSHLINE-30 (wire)
- ❑ TOSHLINE-30 (optical)

I/O Expansion

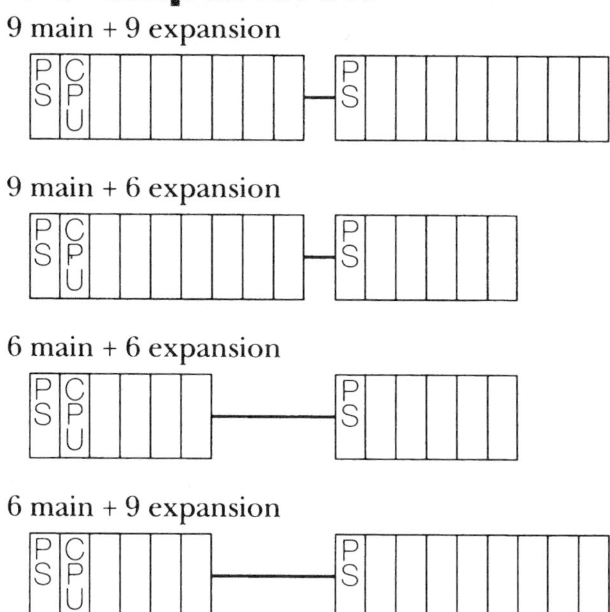

Figure A4.3 Typical input/output expansions

Figure A4.4 Operator's keypad application

EX100 GENERAL SPECIFICATIONS

Power supply voltage
100 to 120 V a.c. (+10%, –15%), 50/60 Hz (±5%)
200 to 240 V a.c. (+10%, –15%), 50/60 Hz (±5%)
24 Vdc (+20%, –15%)

Power consumption
Less than 50 VA (a.c. power supply)
Less than 22 W (d.c. power supply)

Retentive power fault
Less than 10 ms

Withstand voltage
1500 V a.c., 1 minute

Ambient temperature
0 to 55°C for operation
–20 to 75°C for storage

Ambient humidity
20 to 90% RH (no-condensation)

Noise immunity
1000 Vp-p/1 μs, NEMA ICS3–304

Approximate weight
6-slot rack with full modules: less than 2.8 kg
9-slot rack with full modules: less than 4 kg

Peripherals

HANDY PROGRAMMER HP-100

The HP-100 is a hand-held graphic programmer (Fig A4.5). Its portability makes the HP-100 ideal for maintenance use at remote installations, whilst it has all the features of a full-size programming terminal.

- Enter programs in ladder logic
- On-line program edit and monitor (logic intensifies to indicate power flow)
- Block monitor for I/O and internal registers
- On-line data set and I/O force
- Double-length data monitoring/setting

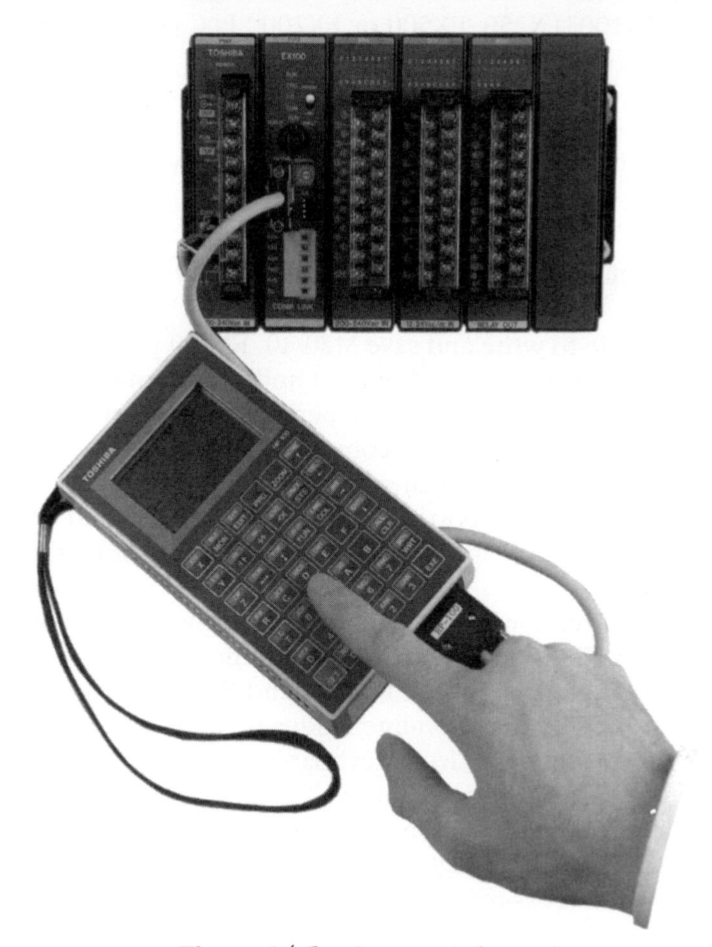

Figure A4.5 Operator's keypad

- ❑ Two display modes
  Full (normal): 5 lines by 11 columns
  Zoom: 2 lines by 2 columns with full device description.

GRAPHIC PROGRAMMER GP110AP1

The GP110 AP1 is an enhanced graphic programmer. It has a large dot matrix LCD screen that displays 7 lines by 11 columns. Logic lines intensify to indicate power flow. Device type, address, current values of timers, counters, and data registers are shown during program execution.

PLC-PLC-REMOTE I/O

The Toshline 30 data link systems enable communication between up to 16 PLC or remote I/O stations on one network (Fig A4.5). It has the following features:

- ❑ Any M20/40, EX100, EX250, EX500 or EX2000 PLC can be utilized
- ❑ Communication over distances of up to 1 km
- ❑ One screened twisted-pair cable only required
- ❑ Very fast transmission (187.5 kpps)
- ❑ Area of memory common to all stations on the network.

EX program development and documentation (EX-PPD™)

Naturally it is possible to write and save M20/40, EX100 programs on a personal computer. The EX program development and documentation software (EX-PDD) runs on any IBM* – PC, XT, AT, PS/2 personal computer and most IBM-PC compatibles such as Toshiba's laptop computers.

- ❑ Built-in modem initialize and dial-up
- ❑ Same EX-PDD software supports M20/40, EX100, EX250 and EX500 PLCs
- ❑ Write ladder/function block programs off-line (PC disk) or on-line (PLC memory)
- ❑ Full-feature ladder editor includes move, copy, insert, delete, search, etc.
- ❑ Make changes in a PLC program while in run mode
- ❑ Load and save programs between PC disk and PLC
- ❑ Monitor power-flow status of on-line ladder program and register values
- ❑ Force I/O and coils on or off from keyboard
- ❑ Document programs with commentary
- ❑ Print ladder program with commentary and in-ladder coil cross-reference
- ❑ Print map options such as register values, instruction usage, device usage, forced devices, full cross-reference, etc.

---

*IBM is a registered trademark of International Business Machines Inc.

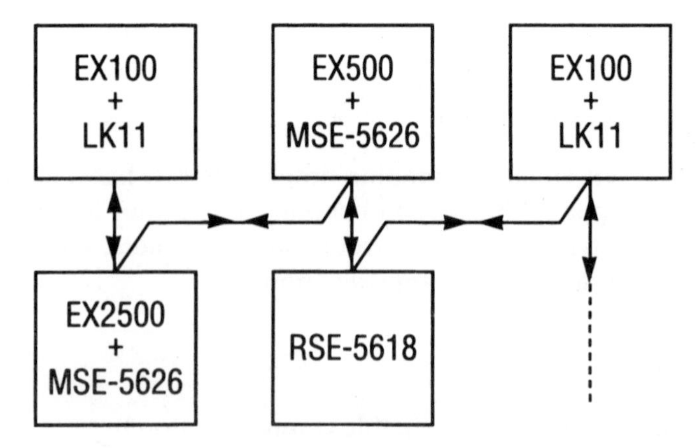

Figure A4.6 Example of system configuration

IPLEX SUPERVISORY SOFTWARE

Iplex is a powerful and flexible scada (system control and data acquisition) system which is used to integrate Toshiba's range of programmable logic controllers to industrial and personal computers. Scada is a powerful software application. The DOS-based system is designed to be extremely 'user friendly' and easy to configure. Standard options include:

- ❑ Real time colour graphics
- ❑ Alarm and event logging
- ❑ Real time and historical trends
- ❑ X and Y bargraph displays, etc.
- ❑ Disk logging
- ❑ Recipe download

Figure A4.7 Graphic screen

Instruction set

Basic ladder functions

Instruction	Expression
NO contact	─┤ ├─ Ⓐ
NC contact	─┤/├─ Ⓐ
Coil	─( )─┤ Ⓐ
Forced coil	✕( )─┤ Ⓐ
Transitional contact (rising)	Input ─┤↑├─ Output Ⓐ
Transitional contact (falling)	Input ─┤↓├─ Output Ⓐ
Master control	Input ── [MCS] ─┤ ≀ ├── [MCR] ─┤
Jump control	Input ── [JCS] ─┤ ≀ ├── [JCR] ─┤
ON delay timer	Input ─[Ⓐ TON Ⓑ]─ Output
OFF delay timer	Input ─[Ⓐ TOF Ⓑ]─ Output
Single-shot timer	Input ─[Ⓐ SS Ⓑ]─ Output
Counter	Count Input ─[CNT Ⓐ Ⓑ]─ Output; Enable Input
End	├──[END]──┤

Data transfer instructions

Instruction (FUN. No.)	Expression
Register transfer (FUN.000)	─[Ⓐ W → W Ⓑ]─
Constant transfer (FUN.001)	─[Ⓐ K → W Ⓑ]─
Table initialization (FUN.002)	─[Ⓐ TINZ [nn] Ⓑ]─
Multiplexer (FUN.003)	─[Ⓐ T → W [nn] Ⓑ → Ⓒ]─
Demultiplexer (FUN.004)	─[Ⓐ W → T [nn] Ⓑ → Ⓒ]─
Table block transfer (FUN.005)	─[Ⓐ T → T [nn] Ⓑ]─

Arithmetic operations

Instruction (FUN. No.)	Expression
Register addition (FUN.010)	─[Ⓐ + Ⓑ → Ⓒ]─
Register subtraction (FUN.011)	─[Ⓐ - Ⓑ → Ⓒ]─
Register multipliction (FUN.012)	─[Ⓐ x Ⓑ → Ⓒ]─
Register division (FUN.013)	─[Ⓐ / Ⓑ → Ⓒ]─
Register comparison (FUN.014)	─[Ⓐ > Ⓑ]─ Output
Register comparison (FUN.015)	─[Ⓐ = Ⓑ]─ Output
Register comparison (FUN.016)	─[Ⓐ < Ⓑ]─ Output

Arithmetic operations

Instruction (FUN. No.)	Expression
Double-length addition (FUN.017)	─[Ⓐ + + Ⓑ → → Ⓒ]─
Double-length subtraction (FUN.018)	─[Ⓐ - - Ⓑ → → Ⓒ]─
Constant addition (FUN.020)	─[Ⓐ +. Ⓑ → Ⓒ]─
Constant subtraction (FUN.021)	─[Ⓐ -. Ⓑ → Ⓒ]─
Constant multiplication (FUN.022)	─[Ⓐ x. Ⓑ → Ⓒ]─
Constant division (FUN.023)	─[Ⓐ /. Ⓑ → Ⓒ]─
Constant comparison (FUN.024)	─[Ⓐ >. Ⓑ]─ Output
Constant comparison (FUN.025)	─[Ⓐ =. Ⓑ]─ Output
Constant comparison (FUN.026)	─[Ⓐ <. Ⓑ]─ Output

Logical operations

Instruction (FUN. No.)	Expression
Register AND (FUN.030)	─[Ⓐ AND Ⓑ → Ⓒ]─
Register OR (FUN.031)	─[Ⓐ OR Ⓑ → Ⓒ]─
Register exclusive OR (FUN.032)	─[Ⓐ EOR Ⓑ → Ⓒ]─
Register inversion (FUN.034)	─[Ⓐ NOT Ⓑ]─

Figure A4.8 Ladder logic instructions

Logical operations

Instruction (FUN. No.)	Expression
Right rotation (FUN.035)	Ⓐ RTR Ⓑ → Ⓒ
Left rotation (FUN.036)	Ⓐ RTL Ⓑ → Ⓒ
Constant AND (FUN.040)	Ⓐ AND. Ⓑ → Ⓒ
Constant OR (FUN.041)	Ⓐ OR. Ⓑ → Ⓒ
Constant exclusive OR (FUN.042)	Ⓐ EOR. Ⓑ → Ⓒ
Bit test (FUN.043)	Ⓐ TEST Ⓑ — Output
Two's Compliment (FUN.046)	Ⓐ NEG Ⓑ

Data conversion instructions

Instruction (FUN. No.)	Expression
Binary conversion (FUN.050)	Ⓐ BIN Ⓑ
Single-length BCD conversion (FUN.051)	Ⓐ BCD1 Ⓑ
Double-length BCD conversion (FUN.052)	Ⓐ BCD2 Ⓑ
Encode (FUN.053)	Ⓐ ENC Ⓑ
Decode (FUN.054)	Ⓐ DEC Ⓑ
Bit count (FUN.055)	Ⓐ BITC Ⓑ

Special functions

Instruction (FUN. No.)	Expression
Upper limit (FUN.060)	Ⓐ UL Ⓑ → Ⓒ
Lower limit (FUN.061)	Ⓐ LL Ⓑ → Ⓒ
Maximum value (FUN.062)	Ⓐ MAX [nn] Ⓑ
Minimum value (FUN.063)	Ⓐ MIN [nn] Ⓑ
Average value (FUN.064)	Ⓐ AVE [nn] Ⓑ
Function generator (FUN.065)	Ⓐ FG [nn] Ⓑ → Ⓒ
Square root (FUN.070)	Ⓐ RT Ⓑ
Sine function (FUN.071)	Ⓐ SIN Ⓑ
Arcsine function (FUN.072)	Ⓐ ASIN Ⓑ
Cosine function (FUN.073)	Ⓐ COS Ⓑ
ArcCosine function (FUN.074)	Ⓐ ACOS Ⓑ

Other functions

Instruction (FUN. No.)	Expression
Device set (FUN.080)	SET Ⓐ
Device reset (FUN.081)	RST Ⓐ
Diagnostic display (FUN.090)	DDSP Ⓐ
Diagnostic display with message (FUN.091)	DDSM Ⓐ Ⓑ
Immediate input (FUN.096)	IN [nn] Ⓐ
Immediate output (FUN.097)	OUT [nn] Ⓐ
Step sequence initialization (FUN.100)	STIZ [nn] Ⓐ
Step sequence input (FUN.101)	Ⓐ
Step sequence output (FUN.102)	Ⓐ
Flip-flop (FUN.110)	Set input, Reset input — F/F Ⓐ
Up/down counter (FUN.111)	Set input, Reset input, Enable input — U/D Ⓐ
Shift register (FUN.112)	Data input, Shift input, Enable input — SR [nn] Ⓐ

Figure A4.9 Ladder logic operations and other functions

APPENDIX 5
GE Fanuc Automation Series 90™ Micro PLC

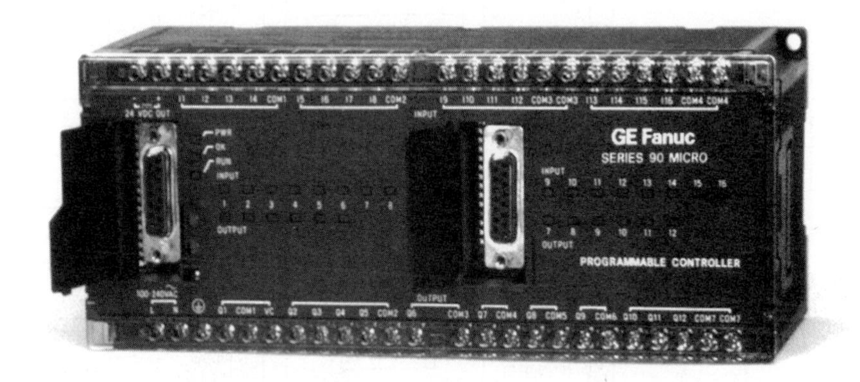

Figure A5.1 Typical PLC units

Programming and Communications Over One Network

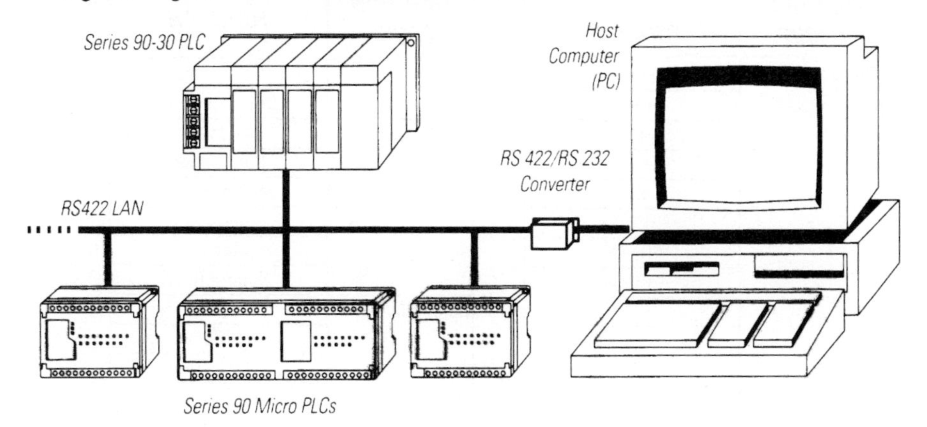

Figure A5.2 A Series 90–30 PLC acting as a master can support up to eight slave Series 90 Micro PLCs

Programming Flexibility

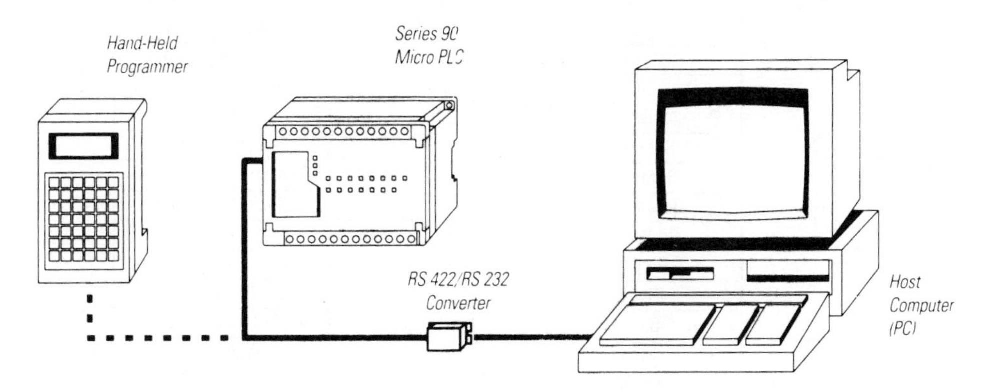

Figure A5.3 Typical system connections

FUNCTION GROUP	FUNCTION
Booleans	--] [--
	--]/[--
	--(↑)--
	--(↓)==
	--(M)--
	--(/M)--
	--(SM)--
	--(RM)--
	--()--
	--(/)--
	--(S)--
	--(R)--
	---<+>
	<+>---
Timers	On Delay Timer
	Off Delay Timer
	Elapsed Timer

1

FUNCTION GROUP	FUNCTION
Counters	Up Counter
	Down Counter
Math	Addition
	Subtraction
	Multiplication
	Division
	Modulo
	Square Root
Relational	Equal
	Not Equal
	Greater Than
	Greater Than/Equal
	Less Than
	Less Than/Equal
	Range
Conversions	Convert To BCD(4)
	Convert To INT

2

FUNCTION GROUP	FUNCTION
Bit Operation	AND
	OR
	Exclusive OR
	NOT
	Shift Left
	Shift Right
	Rotate Left
	Rotate Right
	Bit Position
	Bit Clear
	Bit Test
	Bit Set
Data Move	Move
	Block Move
	Block Clear
	Shift Register
	Bit Sequencer

3

FUNCTION GROUP	FUNCTION
Table	Array Move
	Search Equal
	Search Not Equal
	Search Greater Than
	Search Greater Than/EQ
	Search Less Than
	Search Less Than/EQ
	Masked Compare
Control	DO I/O
	PID - ISA Algorithm
	PID - IND Algorithm
	End Instruction
	Service Request
	Nested MCR
	Nested ENDMCR
	Comm Request

4

Figure A5.4 Examples of the Series 90 basic programming instruction sets. 1 Booleans and Timers; 2 Counters, Math, Relational and Conversions; 3 Bit Operation and Data Move; 4 Table and Control

Series 90 Micro PLC Wiring Diagram Examples

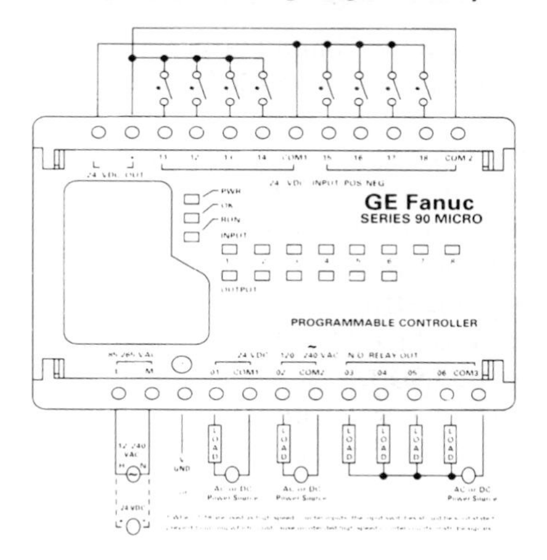

Figure A5.5 Example of Series 90 wiring

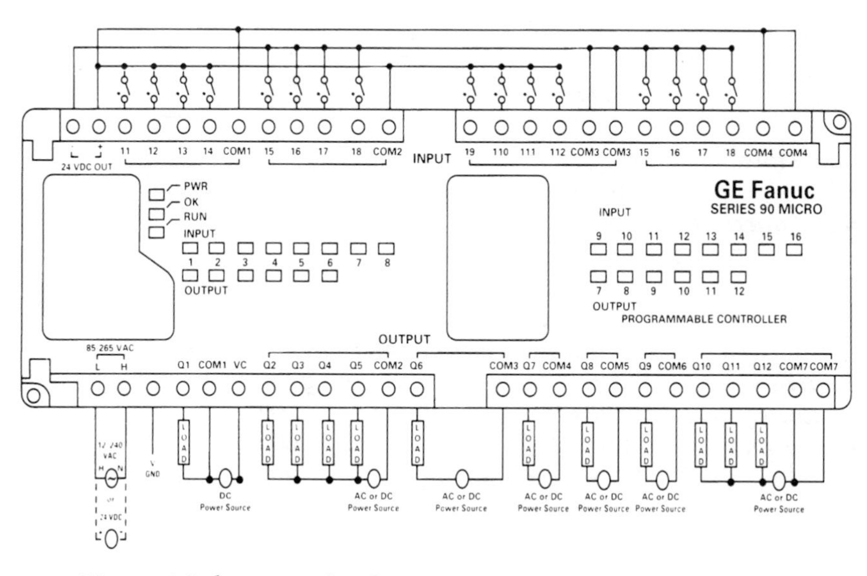

Figure A5.6 Example of Series 90 Micro PLC wiring diagram

General specifications

Dimension (LWH)	115 × 82 × 76mm (14 pt unit) 218 × 82 × 76mm (28 pt unit)
Temperature	
Operating	0 to 55 degrees C
Storage	–40 to 85 degrees C
Mounting Method	35mm DIN Rail or Panel
Vibration	IEC 68.2.6, 2G
Shock	MIL-STD 810C, 15G
Noise Immunity	ANSI/IEEE C37.90A 801.2 ESD L3 801.3 L3, 801.6 RF 801.4 Fast Transient L3 801.5 Surge EN 55011 Emissions
Agency Approvals	FCC [UL (Class 1, Div. 2), CUL and CE pending for certain models]

CPU specifications

	14pt	***28pt***
Execution Time		
Boolean	1.8 ms/K	1.0 ms/K
Typical Function	48 μs	29 μs
Memory		
Size	3 K Words	6 K Words
Type	RAM, Flash EEPROM, Hand-Held Programmer Memory Card	
Data Registers	256	2 K
Internal Coils	1024	1024
Timers/Counters	80	600
Programming	Relay Ladder	Relay Ladder
Instructions	14 Boolean 60+ Advanced	
		2 Ports
Serial Port	RS422, SNP	RS422, SNP RS422, SNP

Answers and Solutions

Chapter 1

1. Sequential control is used to ensure that the correct sequence of events takes place. These are the logical steps required to produce a sequence-controlled program for a PLC system.
2. RAM stands for random access memory. Is a flexible type of read/write memory and is volatile, which means that once the power is removed the data is lost. ROM stands for read-only memory. It is programmed during its manufacture and is a non-volatile memory device which provides a permanent storage for the operating system and fixed data.
3. The EEPROM is used as a back-up memory device and can retain data for long periods of time.
4. CPU stands for central processing unit, which is part of the PLC structure. It is used to perform arithmetic and logical operations on the data in accordance with the instructions stored in memory.
5. An auxiliary relay is an internal device which is a memory element located in RAM. It is called 'auxiliary' because it is an imaginary internal relay.
6. The process by which the PLC gathers information from the inputs, runs the program and then outputs the information in a cyclic fashion.
7. I/O stands for input/output. They are digital input devices and digital output devices interfacing with the PLC.
8. On-line means to enter a new program to the PLC from a file in the computer, or to change data in the registers while the PLC is in run mode.
   Off-line means to enter a new program to a file and save the ladder logic on disk, or to edit, save and verify new programs.
9. A unitary type PLC is a self-contained PLC, i.e. power supply, CPU and a limited number of inputs and outputs.
   A modular type can be constructed using separate modules, i.e. power supply, CPU, input, output, timers, counters, ADC and DAC, motor drives, position control, communications and network modules.
10. The main components of a typical PLC unit are the inputs, CPU, outputs, memory devices, the I/O interfacing and multiplexing and a power supply.

Chapter 2

1. A circuit diagram is a system of conducting parts consisting of switches, relays, transistors, resistors, etc. and their interconnections through which electric current is intended to flow.
   A ladder logic diagram is a series of contacts, timers, counters and coils connected between two vertical supply lines. The left-hand side of the rung contains the inputs and control elements and the output is always located on the right-hand side of the rung.
2. Normally open means that switches and contacts in this state will close when they are operated.
3. Normally closed means that switches and contacts in this state will open when they are operated.
4.

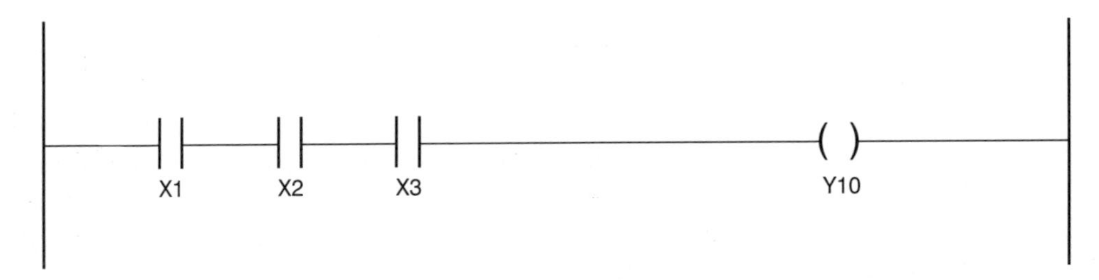

5.

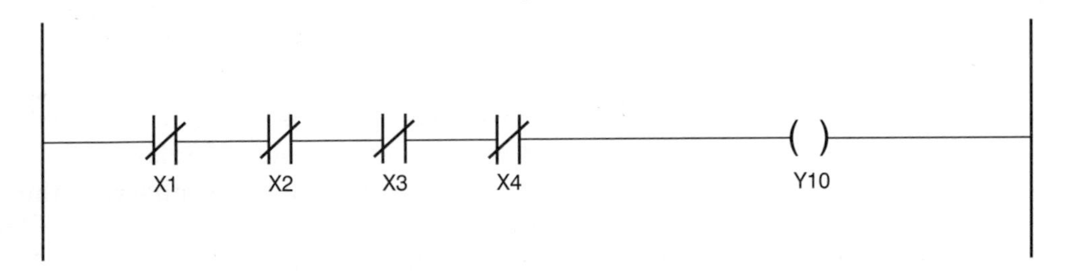

6. The EXCLUSIVE OR gate is to perform the function of non-equivalence, i.e. the output is only obtained when the inputs are not the same.
7. (a) The OR function is diagram (c).
   (b) The AND function is diagram (a).
   (c) The NOR function is diagram (b).
   (d) The NAND function is diagram (d).

Chapter 3

1. Position Number 7 6 5 4 3 2 1 0

Binary	1	1	0	0	1	0	1	0

Denary 128 + 64 + 8 + 2 = 202.

2. 469/2 = 234 $r = 1$
   234/2 = 117 $r = 0$
   117/2 = 58 $r = 1$
   58/2 = 29 $r = 1$
   29/2 = 14 $r = 1$
   14/2 = 7 $r = 0$
   7/2 = 3 $r = 1$
   3/2 = 1 $r = 1$
   1/2 = 0 $r = 1$.
   Therefore the binary number is 1 1 1 0 1 1 1 0 1.
3. C6 = 192 + 6 = 198 in denary
   198/2 = 99 $r = 0$
   99/2 = 49 $r = 1$
   49/2 = 24 $r = 1$
   24/2 = 12 $r = 0$
   12/2 = 6 $r = 0$
   6/2 = 3 $r = 0$
   3/2 = 1 $r = 1$
   1/2 = 0 $r = 1$.

Therefore the binary number is 1 1 0 0 0 1 1 0.

4.

1	2	4	5	Denary
0001	0010	0100	0101	In BCD code.

5.

101	101	Octal
5	5	In Denary.

6.

Position number	7	6	5	4	3	2	1	0
Binary number	0	1	1	1	1	1	1	1

Denary value 64 + 32 + 16 + 8 + 4 + 2 + 1 = 127.
Therefore, answer (d) is correct.

7.

1000	0111	0101
8	7	5

Therefore, answer (c) is correct.

8. A5 = 160 + 5 = 165 in denary
   Therefore, answer (d) is correct.

Chapter 4

1. Correct answer is (b).
2. Correct answer is (c).
3. Correct answer is (b).
4. Correct answer is (c).
5. Correct answer is (b).

6. Correct answer is (b).
7. Solution, where
   X000 is the START button
   X001 is the RESET button
   X00A to X00F are the sensors
   Y010 is the audible alarm

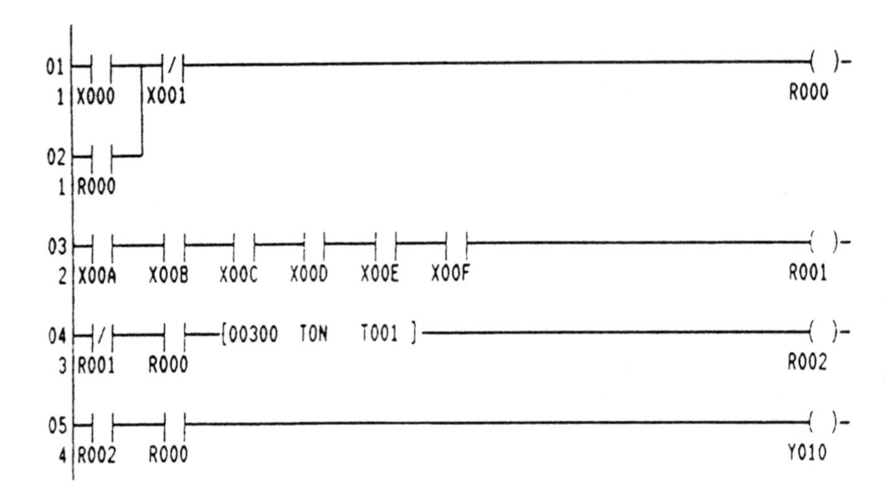

Solution to task NO. 1

```
01 ─┤ ├──┬──┤/├──────────────────────────────────────────( )─
    X000  │  X00F                                          R000
02 ─┤ ├──┘
    R000
03 ─┤ ├──┬──┤/├──────────────────────────────────────────( )─
    X001  │  X00F                                          R00A
04 ─┤ ├──┘
    R00A
05 ─┤ ├────┤/├──[00020   TOF    T004 ]────────────────────( )─
    R000    R00A                                           Y010
06 ─┤ ├──[00010   TON    T001 ]──┤/├──[00010   TOF    T003 ]──( )─
    X00A                         R00A                      Y011
07 ─┤ ├──[00010   TON    T002 ]──┤/├──────────────────────( )─
    X00B                         R00A                      Y012
08 ─┤ ├───────────────────────────────────────────────────( )─
    X00C                                                   R001
```

Solution to task NO. 2

```
01 ─┤ ├──┬──┤/├────┤/├───────────────────────────────────( )─
    X000  │  X001    R003                                  R000
02 ─┤ ├──┘
    R000
03 ─┤ ├────┤/├──[00050   TON    T001 ]────────────────────( )─
    R000    R002                                           R001
04 ─┤ ├──[00050   TON    T002 ]───────────────────────────( )─
    R001                                                   R002
05 ─┤ ├────┤/├────────────────────────────────────────────( )─
    R000    R001                                           Y010
06 ─┤ ├──C   CNT       Q──────────────────────────────────( )─
    R002 |             |                                   R003
07 ─┤ ├──E 00010 C01   |
    R000
```

Solution to task NO. 3

```
01 ├[XW00      BIN   D0100]
02 ├[D0100   - .    00032      >    D0101]
03 ├[D0101   * .    00005    > >    D0103]
04 ├[D0103   / .    00009      >    D0105]
05 ├[D0105   BCD1   D0106]
06 ├[D0106    NOT   YW01  ]
07 ├┤ END ├
```

Chapter 5

1. RS-232C is used for communications over short distances of up to 16 metres. It will only operate with one driver and one receiver. It uses an unbalanced line and for greater distances telephone lines are used with the addition of a modem at each end of the line.
   RS-422A is a balanced digital interface which means that it is isolated from common earth or ground, provides better line matching and can be used for communications over a distance of 1000 metres. It can also be used as a bus system to connect to a number of PLC units. It has a higher transmission rate than RS-232C.
2. Provides the control of the flow of data from a computer to a number of devices and PLCs linked to the communication system and can be defined as the means of transferring digital information in the form of 1's and 0's.
3. The computer can communicate with more than one PLC unit and all the devices can have access to each other and share data within the system.
4. To isolate the low-voltage circuits from the higher-voltage mains supplies; to provide electrical isolation and minimize the possibility of damage to the PLC system, to eliminate ground-loop and earth fault problems, which can cause interference.
5. A diode must always be connected across the relay coil in order to remove the unwanted induced back-emf when the supply to the relay is switched off.
6. See Fig. 5.17. For the transistor sink mode the +24 V supply is made the common connection, whereas for the transistor source mode the 0 V line is made the common connection.

Chapter 6

1. Correct answer is (c).
2. It is a transducer device which converts motion sequences, such as rotation and position, into a digital signal.
3. Twisted-pair cables are used to reduce the effects of electromagnetic interference.
4. Correct answer is (b).
5. The offset represented by 4 mA provides the PLC unit with the distinction between a zero and an open circuit.
6. They can be controlled to operate in an exact number of individual steps for the accurate rotation and position of machines.
7. Correct answer is (b).
8. The resolution will be
   $1/2^{12} = 1/4096 \times 100 = 0.024\%$

Chapter 7

1. Correct answer is (c). It is important to isolate all power when inserting or removing any electrical device, as damage will occur due to the transient currents produced.
2. Correct answer is (b).
3. Emergency stop circuits must be fail-safe. PLCs use solid state devices which are inherently unsafe. Therefore, hard-wired circuits are used and designed to be fail-safe.
4. The earthing of PLCs and associated equipment is necessary to remove the possibility of pick-up noise, due to electromagnetic interference (EMI).
5. Correct answer is (d). The frequency of any service can only be determined by the company after periodic reviews.
6. Correct answer is (d).
7. Correct answer is (c). Although an alarm will sound, the priority is to shut down the machines immediately and safely.

Index